Lecture Notes in Mathematics 1566

Springer
Berlin
Heidelberg
New York
Barcelona
Budapest
Hong Kong
London
Milan
Paris
Santa Clara
Singapore
Tokyo

B. Edixhoven J.-H. Evertse (Eds.)

Diophantine Approximation and Abelian Varieties

Springer

Editors

Bas Edixhoven
Institut Mathématique
Université de Rennes I
Campus de Beaulieu
F-35042 Rennes Cedex, France

Jan-Hendrik Evertse
Universiteit Leiden
Wiskunde en Informatica
Postbus 9512
NL-2300 RA Leiden, The Netherlands

1st edition 1993
2nd printing 1997 (with minor corrections)
3rd printing 2003

Cataloging-in-Publication Data applied for

Die Deutsche Bibliothek - CIP-Einheitsaufnahme

Diophantine approximation and abelian varieties : introductory lectures / B. Edixhoven ; J.-H. Evertse (ed.). - 2. printing. - Berlin ; Heidelberg ; New York ; Barcelona ; Budapest ; Hongkong ; London ; Mailand ; Paris ; Santa Clara ; Singapur ; Tokio : Springer, 1997
(Lecture notes in mathematics ; Vol. 1566)
ISBN 3-540-57528-6

Mathematics Subject Classification (1991): 14G05, 14G40, 11J99, 11G35

ISSN 0075-8434
ISBN 3-540-57528-6 Springer-Verlag Berlin Heidelberg New York

Springer-Verlag Berlin Heidelberg New York
a member of BertelsmannSpringer Science+Businiess Media GmbH

Printed in Germany

Typesetting: Camera-ready TEX output by the author/editor
46/3111-543 - Printed on acid-free paper

Preface

From April 12 to April 16, 1992, the instructional conference for Ph.D-students "Diophantine approximation and abelian varieties" was held in Soesterberg, The Netherlands. The intention of the conference was to give Ph.D-students in number theory and algebraic geometry (but anyone else interested was welcome) some acquaintance with each other's fields. In this conference a proof was presented of Theorem I of G. Faltings's paper "Diophantine approximation on abelian varieties", Ann. Math. 133 (1991), 549–576, together with some background from diophantine approximation and algebraic geometry. These lecture notes consist of modified versions of the lectures given at the conference.

We would like to thank F. Oort and R. Tijdeman for organizing the conference, the speakers for enabling us to publish these notes, C. Faber and W. van der Kallen for help with the typesetting and last but not least the participants for making the conference a successful event.

Contents

Contributors

F. Beukers
Universiteit Utrecht, Mathematisch Instituut, Postbus 80.010, 3508 TA Utrecht, The Netherlands.

B. Edixhoven
Université de Rennes 1, Institut Mathématique, Campus de Beaulieu, 35042 Rennes, France.

J.H. Evertse
Universiteit Leiden, Wiskunde en Informatica, Postbus 9512, 2300 RA Leiden, The Netherlands.

C. Faber
Universiteit van Amsterdam, Wiskunde en Informatica, Plantage Muidergracht 24, 1018 TV Amsterdam, The Netherlands.

G. van der Geer
Universiteit van Amsterdam, Wiskunde en Informatica, Plantage Muidergracht 24, 1018 TV Amsterdam, The Netherlands.

J. Huisman
Université de Rennes 1, Institut Mathématique, Campus de Beaulieu, 35042 Rennes, France.

A.J. de Jong
Princeton University, Fine Hall, Princeton, NJ 08544-0001, U.S.A.

R.J. Kooman
Lijtweg 607, 2341 Oegstgeest, The Netherlands.

F. Oort
Universiteit Utrecht, Mathematisch Instituut, Postbus 80.010, 3508 TA Utrecht, The Netherlands.

M. van der Put
Universiteit Groningen, Wiskunde en Informatica, Postbus 800, 9700 AV Groningen, The Netherlands.

R. Tijdeman
Universiteit Leiden, Wiskunde en Informatica, Postbus 9512, 2300 RA Leiden, The Netherlands.

J. Top
Universiteit Groningen, Wiskunde en Informatica, Postbus 800, 9700 AV Groningen, The Netherlands.

Introduction

Although diophantine approximation and algebraic geometry have different roots, today there is a close interaction between these fields. Originally, diophantine approximation was the branch in number theory in which one deals with problems such as approximation of irrational numbers by rational numbers, transcendence problems such as the transcendence of e or π, etc. There are some very powerful theorems in diophantine approximation with many applications, among others to certain classes of diophantine equations. It turned out that several results from diophantine approximation could be improved or generalized by techniques from algebraic geometry. The results from diophantine approximation which we discuss in detail in these lecture notes are Roth's theorem, which states that for every algebraic number α and for every $\delta > 0$ there are only finitely many $p, q \in \mathbb{Z}$ with $|\alpha - p/q| < |q|^{-2-\delta}$, and a powerful higher dimensional generalization of this, the so-called Subspace theorem of W.M. Schmidt. Here, we would like to mention the following consequence of the Subspace theorem, conjectured by S. Lang and proved by M. Laurent: let Γ be the algebraic group $(\overline{\mathbb{Q}}^*)^n$, endowed with coordinatewise multiplication, V a subvariety of Γ, not containing a translate of a positive dimensional algebraic subgroup of Γ, and G a finitely generated subgroup of Γ; then $V \cap \Gamma$ is finite.

We give a brief overview of the proof of Roth's theorem. Suppose that the equation $|\alpha - p/q| < q^{-2-\delta}$ has infinitely many solutions $p, q \in \mathbb{Z}$ with $q > 0$. First one shows that for sufficiently large m there is a polynomial $P(X_1, \ldots, X_m)$ in $\mathbb{Z}[X_1, \ldots, X_m]$ with "small" coefficients and vanishing with high order at $(\alpha, \ldots, \alpha)$. Then one shows that P cannot vanish with high order at a given rational point $x = (p_1/q_1, \ldots, p_n/q_n)$ satisfying certain conditions. This non-vanishing result, called Roth's Lemma, is the most difficult part of the proof. From the fact that $|\alpha - p/q| < q^{-2-\delta}$ has infinitely many solutions it follows that one can choose x such that $|\alpha - p_n/q_n| < q_n^{-2-\delta}$ for n in $\{1, \ldots, m\}$. Then for some small order partial derivative P_i of P we have $P_i(x) \neq 0$. But $P_i(x)$ is a rational number with denominator dividing $a := q_1^{d_1} \cdots q_m^{d_m}$, where $d_j = \deg_{X_j}(P_i)$. Hence $|P_i(x)| \geq 1/a$. On the other hand, P_i is divisible by a high power of $X_j - \alpha$ and $|p_j/q_j - \alpha|$ is small for all j in $\{1, \ldots, m\}$. Hence $P_i(x)$ must be small. One shows that in fact $|P_i(x)| < 1/a$ and thus one arrives at a contradiction.

Algebraic geometry enables one to study the geometry of the set of solutions (e.g., over an algebraically closed field) of a set of algebraic equations. The geometry often predicts the structure of the set of arithmetic solutions (e.g., over a number field) of these algebraic equations. As an example one can mention Mordell's conjecture, which was proved by G. Faltings in 1983 [21]. Several results of this type have been proved by combining techniques from algebraic geometry with techniques similar to those used in the proof of Roth's theorem. Typical examples are the Siegel-Mahler finiteness theorem for integral points on algebraic curves and P. Vojta's recent proof of Mordell's conjecture.

In these lecture notes, we study the proof of the following theorem of G. Faltings ([22], Thm. I), which is the analogue for abelian varieties of the result for $(\overline{\mathbb{Q}}^*)^n$

mentioned above, and which was conjectured by S. Lang and by A. Weil:

> *Let A be an abelian variety over a number field k and let X be a subvariety of A which, over some algebraic closure of k, does not contain any positive dimensional abelian variety. Then the set of rational points of X is finite.*

(Note that this theorem is a generalization of Mordell's conjecture.) The proof of Faltings is a higher dimensional generalization of Vojta's proof of Mordell's conjecture and has some similarities with the proof of Roth's theorem. Basically it goes as follows. Assume that $X(k)$ is infinite. First of all one fixes a very ample symmetric line bundle $\mathcal{L}$ on A, and norms on $\mathcal{L}$ at the archimedian places of k. Let m be a sufficiently large integer. There exists $x = (x_1, \ldots, x_m)$ in $X^m(k)$ satisfying certain conditions (e.g., the angles between the x_i with respect to the Néron-Tate height associated to $\mathcal{L}$ should be small, the quotient of the height of x_{i+1} by the height of x_i should be big for $1 \leq i < m$ and the height of x_1 should be big). Instead of a polynomial one then constructs a global section f of a certain line bundle $\mathcal{L}(\sigma - \varepsilon, s_1, \ldots, s_m)^d$ on a certain model of X^m over the ring of integers R of k. This line bundle is a tensor product of pullbacks of $\mathcal{L}$ along maps $A^m \to A$ depending on $\sigma - \varepsilon$, the s_i and on d; in particular, it comes with norms at the archimedian places. By construction, f has small order of vanishing at x and has suitably bounded norms at the archimedian places of k. Then one considers the Arakelov degree of the metrized line bundle $x^*\mathcal{L}(\sigma - \varepsilon, s_1, \ldots, s_m)^d$ on $\mathrm{Spec}(R)$; the conditions satisfied by the x_i give an upper bound, whereas the bound on the norm of f at the archimedian places gives a lower bound. It turns out that one can choose the parameters ε, σ, the s_i and d in such a way that the upper bound is smaller than the lower bound.

We mention that the construction of f is quite involved. Intersection theory is used to show that under suitable hypotheses, the line bundles $\mathcal{L}(-\varepsilon, s_1, \ldots, s_m)^d$ are ample on X^m. A new, basic tool here is the so-called Product theorem, a strong generalization by Faltings of Roth's Lemma.

On the other hand, Faltings's proof of Thm. I above is quite elementary when compared to his original proof of Mordell's conjecture. For example, no moduli spaces and no l-adic representations are needed. Also, the proof of Thm. I does not use Arakelov intersection theory. Faltings's proof of Thm. I in [22] seems to use some of it, but that is easily avoided. The Arakelov intersection theory in [22] plays an essential role in the proof of Thm. II of [22], where one needs the notion of height not only for points but for subvarieties; we do not give details of that proof. The only intersection theory that we need concerns intersection numbers obtained by intersecting closed subvarieties of projective varieties with Cartier divisors, so one does not need the construction of Chow rings. The deepest result in intersection theory needed in these notes is Kleiman's theorem stating that the ample cone is the interior of the pseudo-ample cone. Unfortunately, we will have to use the existence and quasi-projectivity of the Néron model over $\mathrm{Spec}(R)$ of A in the proof of Lemma 3.1 of Chapter XI; a proof of that lemma avoiding the use of Néron models would significantly simplify the proof of Thm. I. We believe that for someone with a basic knowledge of algebraic geometry, say Chapters II and III of [27], everything in these notes except for the use of Néron models is not hard to understand. In the case where X is a curve, i.e., Mordell's conjecture, the proof of Thm. I can be considerably simplified; this was done by E. Bombieri in [9].

Let us now describe the contents of the various chapters. Chapter I gives an overview of several results and conjectures in diophantine approximation and arithmetic geometry. After that, the lecture notes can be divided in three parts.

The first of these parts consists of Chapters II–IV; some of the most important results from diophantine approximation are discussed and proofs are sketched of Roth's theorem and of the Subspace theorem.

The second part, which consists of Chapters V–XI, deals with the proof of Thm. I above. Chapters V and VII provide the results needed of the theory of height functions and of intersection theory, respectively. Chapter VIII contains a proof of the Product theorem. This theorem is then used in Chapter IX in order to prove the ampleness of certain $\mathcal{L}(-\varepsilon, s_1, \ldots, s_m)^d$. Chapter X gives a proof of Faltings's version of Siegel's Lemma. Chapter XI finally completes the proof of Thm. I. Chapter VI gives some historical background on how D. Mumford's result on the "widely spacedness" of rational points of a curve of genus at least two over a number field lead to Vojta's proof of Mordell's conjecture.

The third part consists of Chapters XII and XIII. Chapter XII gives an application of Thm. I to the study of points of degree d on curves over number fields. Chapter XIII discusses a generalization by Faltings of Thm. I, which was also conjectured by Lang.

Terminology and Prerequisites

In these notes it will be assumed that the reader is familiar with the basic objects of elementary algebraic number theory, such as the ring of integers of a number field, its localizations and completions at its maximal ideals, and the various embeddings in the field of complex numbers. The same goes more or less for algebraic geometry. To understand the proof of Faltings's Thm. I the reader should be familiar with schemes, morphisms between schemes and cohomology of quasi-coherent sheaves of modules on schemes. In order to encourage the reader, we want to mention that Hartshorne's book [27], especially Chapters II, §§1–8 and III, §§1–5 and §§8–10, contains almost all we need. The two most important exceptions are Kleiman's theorem on the ample and the pseudo-ample cones (see Chapter VII), for which one is referred to [28], and the existence and quasi-projectivity of Néron models of abelian varieties (used in Chapter XI), for which [11] is an excellent reference. At a few places the "GAGA principle" (see [27], Appendix B) and some algebraic topology of complex analytic varieties are used. A less important exception is the theorem of Mordell-Weil, a proof of which can for example be found in Manin's [52], Appendix II, or in [70]; Chapter V of these notes contains the required results on heights on abelian varieties. Almost no knowledge concerning abelian varieties will be assumed. By definition an abelian variety over a field k will be a commutative projective connected algebraic group over k. We will use that the associated complex analytic variety of an abelian variety over $\mathbb{C}$ is a complex torus.

Since these notes are written by various authors, the terminologies used in the various chapters are not completely the same. For example, Chapter I uses a normalization of the absolute values on a number field which is different from the normalization used by the other contributors; the reason for this normalization in Chapter I is clear, since one no longer has to divide by the degree of the number field in question to define the absolute height, but it has the disadvantage that the absolute value no longer just depends on the completion of the number field with respect to the absolute value. Another example is the notion of variety. If k is a field, then by a *(algebraic) variety (defined) over k* one can mean an integral, separated k-scheme of finite type; but one can also mean the following: an *(absolutely irreducible) affine variety (defined) over k* is an irreducible Zariski closed subset in some affine space K^n (K a fixed algebraically closed field containing k) defined by polynomials with coefficients in k, and a *(absolutely irreducible) variety (defined) over k* is an object obtained by glueing affine varieties over k with respect to glueing data given again by polynomials with coefficients in k. As these two notions are (supposed to be) equivalent, no (serious) confusion should arise.

Chapter I

Diophantine Equations and Approximation

by Frits Beukers

1 Heights

Let F be an algebraic number field. The set of valuations on F is denoted by M_F. Let $|.|_v$, or v in shorthand, be a valuation of F. Denote by F_v the completion of F with respect to v. If F_v is $\mathbb{R}$ or $\mathbb{C}$ we assume that v coincides with the usual absolute value on these fields. When v is a finite valuation we assume it normalised by $|p|_v = 1/p$ where p is the unique rational prime such that $|p|_v < 1$. The *normalised* valuation $||.||_v$ is defined by

$$||x||_v = |x|_v^{[F_v:\mathbb{Q}_p]/[F:\mathbb{Q}]}$$

with the convention that $p = \infty$ when v is archimedean and $\mathbb{Q}_\infty = \mathbb{R}$. For any non-zero $x \in F$ we have the *product formula*

$$\prod_v ||x||_v = 1. \tag{1.1}$$

Let L be any finite extension of F. Then any valuation w of L restricted to F is a valuation v of F. We have for any $x \in F$ and $v \in M_F$,

$$||x||_v = \prod_{w|v} ||x||_w \tag{1.2}$$

where the product is over all valuations $w \in M_L$ whose restriction to F is v. The *absolute multiplicative height* of x is defined by

$$H(x) = \prod_v \max(1, ||x||_v).$$

It is a consequence of (1.2) that $H(x)$ is independent of the field F which contains x. The *absolute logarithmic height* is defined by

$$h(x) = \log H(x).$$

Let $\mathbb{P}^n$ be the n-dimensional projective space and let $P \in \mathbb{P}^n(F)$ be an F-rational point with homogeneous coordinates $(x_0, x_1, \ldots, x_n)$. We define the *projective (absolute) height* by

$$h(P) = \sum_v \log \max(||x_0||_v, ||x_1||_v, \ldots, ||x_n||_v).$$

Again, $h(P)$ is independent of the field F containing P. Therefore the projective height can be considered as a function on $\mathbb{P}^n(\overline{F})$. Notice that the height $h(x)$ of a number coincides with the projective height of the point $(1 : x) \in \mathbb{P}^1$. The projective height has the fundamental property that, given h_0, there are only finitely many $P \in \mathbb{P}^n(F)$ such that $h(P) < h_0$.

Let V be a non-singular projective variety defined over F. Let $\phi : V \hookrightarrow \mathbb{P}^N$ be a projective embedding also defined over F. On $V(\overline{F})$, the $\overline{F}$-rational points of V, we take the restriction of the projective height as a height function and denote it by h_ϕ. In general the construction of heights on V runs as follows. First, let D be a very ample divisor. That is, letting $f_0, f_1, \ldots, f_n$ be a basis of the space of all rational functions f defined over F with $(f) \geq -D$, the map $\phi : V \to \mathbb{P}^n$ given by $P \mapsto (f_0(P), f_1(P), \ldots, f_n(P))$ is a projective embedding. The height h_D is then simply defined as h_ϕ. If D_1, D_2 are two linearly equivalent very ample divisors, then $h_{D_1} - h_{D_2}$ is known to be a bounded function on $V(\overline{F})$.

Now let D be any divisor. On a non-singular projective variety one can always find two very ample divisors X, Y such that $D + Y = X$. Define $h_D = h_X - h_Y$. Again, up to a bounded fuction, h_D is independent of the choice of X and Y.

We summarize this height construction as follows.

1.3 Theorem. *There exists a unique homomorphism*

$$\text{linear divisor classes} \quad \to \quad \begin{array}{c}\text{real valued functions on } V(\overline{F}) \\ \text{modulo bounded functions}\end{array}$$

denoted by $c \mapsto h_c + O(1)$ such that: if c contains a very ample divisor, then h_c is equivalent to the height associated with a projective embedding obtained from the linear system of that divisor.

We also recall the following theorem.

1.4 Theorem. *Let c be a linear divisor class which contains a positive divisor Z. Then*

$$h_c(P) \geq O(1)$$

for all $P \in V(\overline{F})$, $P \notin supp(Z)$.

For the proof of the two above theorems we refer to Lang's book [36], Chapter 4.

Finally, following Lang, we introduce the notion of *pseudo ample divisor*, not to be confused with the pseudo ample cone. A divisor D on a variety V is said to be pseudo ample if some multiple of D generates an embedding from some non-empty Zariski open part of V into a locally closed part of projective space. One easily sees that there exists a proper closed subvariety W of V such that, given h_0, the inequality $h_D(P) < h_0$ has only finitely many solutions in $V(F) - W$.

2 The Subspace Theorem

For the sake of later comparisons we shall first state the so-called *Liouville inequality.*

2.1 Theorem (Liouville). *Let F be an algebraic number field and L a finite extension. Let S be a finite set of valuations and extend each $v \in S$ to L. Then, for every $\alpha \in L$, $\alpha \neq 0$ we have*

$$\prod_{v \in S} ||\alpha||_v \geq \frac{1}{H(\alpha)^{[L:F]}}.$$

Proof. Let us assume that $||\alpha||_v < 1$ for every $v \in S$. If not, we simply reduce the set S. Let S_L be the finite set of valuations on L which are chosen as extension of v on F. Using the product formula we find that

$$\begin{aligned}\prod_{w \in S_L} ||\alpha||_w &= \prod_{w \notin S_L} ||\alpha||_w^{-1} \\ &\geq \prod_{w \notin S_L} \max(1, ||\alpha||_w)^{-1} \\ &\geq \prod_{w \in M_L} \max(1, ||\alpha||_w)^{-1} = \frac{1}{H(\alpha)}.\end{aligned}$$

The proof is finished by noticing that

$$||\alpha||_v = ||\alpha||_w^{[L:F]/[L_w:F_v]} \geq ||\alpha||_w^{[L:F]}$$

□

Liouville applied more primitive forms of this inequality to obtain lower bounds for the approximation of fixed algebraic numbers by rationals. In our more general setting, let α be a fixed algebraic number of degree d over F. Then it is a direct consequence of the previous theorem that

$$\prod_{v \in S} (||x - \alpha||_v) > \frac{c(\alpha)}{H(x)^d} \tag{2.2}$$

for every $x \in F$ with $x \neq 0$. Here $c(\alpha)$ is a constant which can be taken to be $(2H(\alpha))^{-d}$. Using such an inequality Liouville was the first to prove the existence of transcendental numbers by constructing numbers which could be approximated by rationals much faster than algebraic numbers. In 1909 A. Thue provided the first non-trivial improvement over (2.2) which was subsequently improved by C.L. Siegel (1921), F. Dyson (1948) and which finally culminated in Roth's theorem, proved around 1955. The theorem we state here is a version by S. Lang which includes non-archimedean valuations, first observed by Ridout, and a product over different valuations.

2.3 Theorem (Roth). *Let F be an algebraic number field and S a finite set of valuations of F. Let $\epsilon > 0$. Let $\alpha \in \overline{\mathbf{Q}}$ and extend each v to $F(\alpha)$. Then*

$$\prod_{v \in S} (||x - \alpha||_v) < \frac{1}{H(x)^{2+\epsilon}}$$

has only finitely many solutions $x \in F$.

A proof of Roth's original theorem can be found in Chapter III of these notes. Around 1970 W.M. Schmidt extended Roth's techniques in a profound way to obtain a simultaneous approximation result. Again the version we state here is a later version which follows from work of H.P. Schlickewei.

2.4 Theorem (Schmidt's Subspace theorem). *Let F be an algebraic number field and S a finite set of valuations of F. Let $H_1, H_2, \ldots, H_m$ be m hyperplanes of $\mathbb{P}^n$ in general position. For each i let H_i be given by $L_i(x_0, \ldots, x_n) = 0$ where L_i is a linear form with coefficients in $\overline{\mathbb{Q}}$. Extend each $v \in S$ to the field generated by the coefficients of the L_i. Let $\epsilon > 0$. Then the points $P = (x_0, \ldots, x_n) \in \mathbb{P}^n(F)$ which satisfy*

$$\prod_{v \in S} \prod_{i=1}^{m} \min_j \left\| \frac{L_i(x_0, \ldots, x_n)}{x_j} \right\|_v < \frac{1}{H(P)^{n+1+\epsilon}}$$

lie in a finite union of proper hyperplanes of $\mathbb{P}^n$.

For a rough sketch of the proof of Schmidt's original theorem we refer to Chapter IV.

Let us rewrite these theorems, starting with Roth's theorem. Take logarithms. Then we find that

$$\sum_{v \in S} \log ||x - \alpha||_v < -(2 + \epsilon) h(x)$$

has finitely many solutions. The function on the left can be considered as a distance function, measuring the S-adic distance between x and α. Minus this function can be considered as a *proximity function* which becomes larger as x gets closer to α. Roth's theorem can now be reformulated as follows. For all but finitely many $x \in F$ we have

$$-\sum_{v \in S} \log ||x - \alpha||_v < (2 + \epsilon) h(x).$$

We can reformulate the Subspace theorem similarly. There is a finite union Z of hyperplanes in $\mathbb{P}^n$ such that for all $P \in \mathbb{P}^n(F)$, $P \notin Z$, we have

$$\sum_{v \in S} \log \max_j \left\| \prod_{i=1}^{m} \frac{x_j}{L_i(x_0, \ldots, x_n)} \right\|_v < (n + 1 + \epsilon) h(P).$$

The left hand side of this inequality can be considered as a proximity function measuring the S-adic closeness of the point P to the union of hyperplanes $\cup_{i=1}^m H_i$.

For comparison, under the same assumptions a trivial application of the Liouville inequality would give that for all $P \in \mathbb{P}^n(F)$, $P \notin H_i$ $(i = 1, \ldots, m)$,

$$\sum_{v \in S} \log \max_j \left\| \prod_{i=1}^{m} \frac{x_j}{L_i(x_0, \ldots, x_n)} \right\|_v \leq [L : F] m h(P) + O(1)$$

where L is the field generated over F by the coefficients of the L_i.

As a comment we would like to add that the Subspace theorem, in the way we have formulated it here, has two crucially distinct ranges of applicability. The first which we call the arithmetic one is when $m \leq n + 1$. The Subspace theorem gives a non-trivial result as long as $[L : F] m > n + 1$. However if we allow points P in some finite extension of F it may happen that $[L : F]$ decreases and the statement of the theorem becomes trivial. The second range is when $m \geq n + 2$, which we like to call the geometric realm. Here it does not matter if we extend the field F, the theorem remains non-trivial.

3 Weil Functions

On general varieties we can also define proximity functions. In that case they are known as *Weil functions.* The general definition is too cumbersome to state here, and we refer to [36], Chapter 10. Instead we give a short recipe for the construction of Weil functions on projective varieties. First construct sets of positive divisors X_i, $(i = 1, \ldots, n)$ and Y_j, $(j = 1, \ldots, m)$ such that $D + X_i \sim Y_j$ for every i, j and such that the X_i have no point in common and the Y_j have no point in common. Let $f_{ij} \in \overline{F}(V)$ be such that $(f_{ij}) = Y_j - X_i - D$ for each pair i, j. Extend the valuation to all of $\overline{F}$. For each $P \in V(\overline{F})$ we define

$$\lambda_{D,v}(P) = \max_j \min_i \log \|f_{ij}(P)\|_v.$$

Of course $\lambda_{D,v}$ depends upon the choice of the f_{ij}, but it is known that two Weil functions associated to the same divisor D differ only by a bounded function on $V(\overline{F})$. For more details on Weil functions we refer to Lang's book [36], Chapter 10.

Consider for example the elliptic curve in $\mathbb{P}^2$ with affine equation

$$y^2 = x^3 + Ax + B, \qquad A, B \in F.$$

Let D be three times the point at infinity. Choose $f_{11} = x$, $f_{12} = y$ and $f_{13} = 1$. Notice that $(x) = \{\infty, (0, \pm\sqrt{B})\} - D$ and $(y) = \{\text{2-torsion} \neq \infty\} - D$. Notice that when $B \neq 0$ the zero divisors of x and y are disjoint. The corresponding Weil function is

$$\lambda_{D,v} = \log \max(1, \|x\|_v, \|y\|_v).$$

As another example take for D the union of hyperplanes $\cup_{i=1}^m H_i$ in $\mathbb{P}^n$ from the Subspace theorem. Let L be the product of all L_i. For the functions f_{ij} we choose

$$f_{1j} = \frac{x_j^m}{L(x_0, \ldots, x_n)} \qquad j = 0, \ldots, n.$$

The pole divisor of each f_{1j} is precisely D and the zero divisor is m times the hyperplane $x_j = 0$. Our v-adic proximity function (Weil function) reads

$$\lambda_{D,v}(P) = \log \max_i \left\| \frac{x_i^m}{L(x_0, \ldots, x_n)} \right\|_v.$$

The Subspace theorem can again be reformulated. There is a finite union of hyperplanes Z in $\mathbb{P}^n$ such that for all $P \in \mathbb{P}^n(F)$, $P \notin Z$ we have

$$\sum_{v \in S} \lambda_{D,v}(P) < (n + 1 + \epsilon) h(P).$$

4 Vojta's Conjecture

In the beginning of the 1980's P. Vojta [80] discovered an uncanny similarity between concepts from diophantine approximation and from value distribution theory for complex analytic functions. Theorems in the latter area, translated via Vojta's dictionary to diophantine approximation, yielded statements which can be considered as very striking conjectures. Although we are far from being able to prove these conjectures, they look like a fascinating guide line in further development of diophantine approximation and diophantine equations. Here we state only Vojta's "Main Conjecture" without going into the analogy with value distribution theory. A *normal crossing divisor* D of a non-singular variety V is a divisor which has a local equation of the form $z_1 z_2 \cdots z_m = 0$ near every point of D for suitably chosen local coordinates $z_1, z_2, \ldots, z_{\dim(V)}$ on V.

4.1 Conjecture (Vojta). *Let V be a non-singular projective variety defined over the algebraic number field F and let A be a pseudo ample divisor. Let D be a normal crossing divisor defined over a finite extension of F. Let K be the canonical divisor of V. Let S be a finite set of valuations on F and for each $v \in S$ let $\lambda_{D,v}$ be a proximity function for D. Let $\epsilon > 0$. Then there exists a Zariski closed subvariety Z of V such that for all $P \in V(F)$, $P \notin Z$ we have,*

$$\sum_{v \in S} \lambda_{D,v}(P) + h_K(P) < \epsilon h_A(P) + O(1).$$

The subvariety Z will be referred to as an *exceptional subvariety*. Actually, an exceptional subvariety is nothing but a Zariski closed subvariety, but it sounds more suggestive for this occasion. One could argue whether or not the condition "non-singular" is really necessary. The "normal crossings" condition is vital however. In the example of the Subspace theorem this condition comes down to the condition that the hyperplanes lie in general position.

As an example consider $V = \mathbb{P}^n$. To determine the canonical divisor consider the differential form $\Omega = dx_1 dx_2 \cdots dx_n$ in the affine coordinates $(1 : x_1 : x_2 : \ldots : x_n)$ on $U_0 = \{x \in \mathbb{P}^n | x_0 \neq 0\}$. There are no zeros or poles on U_0. But if we rewrite Ω with respect to $(x_0 : \ldots : 1 : \ldots x_n)$ on $U_i = \{x \in \mathbb{P}^n | x_i \neq 0\}$ we find

$$\Omega = -\frac{1}{x_0^{n+1}} dx_0 \ldots d\check{x}_i \ldots dx_n.$$

Hence Ω has a pole of order $n+1$ along $x_0 = 0$ and $K = -(n+1)H$ where H is the hyperplane at infinity. So by linearity of heights we find that $h_K = -(n+1)h$ where h is the ordinary projective height. Thus Vojta's conjecture for $V = \mathbb{P}^n$ reads as follows. For all $P \in \mathbb{P}(F)$, $P \notin Z$ we have

$$\sum_{v \in S} \lambda_{D,v}(P) < (n+1+\epsilon)h(P) + O(1).$$

So in particular, when D is a union of hyperplanes in general position we recover the Subspace theorem again, except that Z is now known to be a union of hyperplanes. If we take for D any hypersurface and, to fix ideas, $F = \mathbb{Q}$, we obtain the following interesting consequence.

4.2 Conjecture. *Let D be a hypersurface in $\mathbb{P}^n$ defined over $\mathbb{Q}$ of degree $d \geq n+2$ with at most normally crossing singularities. Suppose it is given by the homogeneous equation $Q(x_0, \ldots, x_n) = 0$ where Q has coefficients in $\mathbb{Z}$. Let S be a finite set (possibly empty) of rational primes. Then the set of points $(x_0, \ldots, x_n) \in \mathbb{Z}^{n+1}$, $\gcd(x_0, \ldots, x_n) = 1$ such that $Q(x_0, \ldots, x_n)$ only contains primes from S lies in a Zariski closed subset of $\mathbb{P}^n$.*

Proof. We apply Vojta's conjecture with the proximity functions

$$\lambda_{D,v} = \log \max_i \left\| \frac{x_i^d}{Q(x_0,\ldots,x_n)} \right\|_v$$

and the set of valuations $S \cup \infty$. The conjecture implies that for any $\epsilon > 0$ the set of projective $n+1$-tuples $(x_0,\ldots,x_n) \in \mathbb{Z}^{n+1}$ with $\gcd(x_0,\ldots,x_n) = 1$ which satisfy

$$\sum_{v \in S \cup \infty} \log \max_i \left\| \frac{x_i^d}{Q(x_0,\ldots,x_n)} \right\|_v > (n+1+\epsilon) h(x_0,\ldots,x_n)$$

lies in an exceptional subvariety. The inequality can be restated as

$$\sum_{v \in S \cup \infty} \log \|Q(x_0,\ldots,x_n)\|_v < -(n+1+\epsilon)h(x_0,\ldots,x_n) + \sum_{v \in S \cup \infty} \log \max_i (\|x_i^d\|_v)$$

The sum on the right is precisely $dh(x_0,\ldots,x_n)$ since the sum includes the infinite valuation and the $\max_i$ is 1 for all finite v. So the set of solutions to

$$\sum_{v \in S \cup \infty} \log \|Q(x_0,\ldots,x_n)\|_v < (d-n-1-\epsilon) h(x_0,\ldots,x_n)$$

lies in an exceptional subvariety. Suppose $Q(x_0,\ldots,x_n)$ is composed of primes only from S. Then by the product formula the left hand side of the latter inequality is zero and we have a solution of the inequality because $d \geq n+1$. This proves our corollary. □

Application of this corollary to the case $n = 1$ yields the *Thue-Mahler equation* $F(x,y) = p_1^{k_1} \cdots p_s^{k_s}$ where F is a binary form of degree at least 3 and distinct zeros. Application of the corollary to $Q = x_0 x_1 \cdots x_n (x_0 + x_1 + \cdots + x_n)$ gives us the S-unit equation in $n+2$ variables.

Application of the corollary in the case $n = 2$ already poses us with problems that no one knows how to solve. We obtain a so-called *ternary form equation*

$$Q(x,y,z) = p_1^{k_1} \cdots p_r^{k_r} \tag{4.3}$$

where Q is a ternary form of degree d and $p_1,\ldots,p_r$ are given primes. A number which is composed of only the primes $p_1,\ldots,p_r$ will be called an S-unit. We assume that the curve $Q(x,y,z) = 0$ has at most simple singularities. Let us distinguish three cases, the case $d = 1$ being considered trivial.

$d = 2$ Here it is very easy to construct examples having a Zariski dense set of solutions. Suppose that Q is indefinite and that there are integers x_0, y_0, z_0 such that $Q(x_0, y_0, z_0) = 1$. We indicate how this implies the existence of a Zariski dense (in $\mathbb{P}^2$) set of solutions to $Q(x,y,z) = 1$, $x,y,z \in \mathbb{Z}$. Take any triple of integers $x_1, y_1, z_1 \in \mathbb{Z}$ and consider the binary form $Q(\lambda x_0 + \mu x_1, \lambda y_0 + \mu y_1, \lambda z_0 + \mu z_1)$ which has the shape $\lambda^2 + A\lambda\mu + B\mu^2$. Because Q is indefinite we can choose x_1, y_1, z_1 in infinitely many ways such that $A^2 - 4B$ is positive and not a square. By writing down an infinite set of solutions λ, μ of the Pellian equation $\lambda^2 + A\lambda\mu + B\mu^2 = 1$ for each such x_1, y_1, z_1 we arrive at our dense set of solutions.

$d=3$ This case seems to be very interesting. When $Q=0$ has a singular point it is often possible to construct a dense set of solutions to (4.3) by considering straight lines through the singularity in a way similar to the previous case. When $Q=0$ is a non-singular curve, things seem to be more difficult, but we can still give examples of dense solution sets. The equation $x^3+y^3+z^3=1$, for example, has a Zariski dense set of solutions $x,y,z\in\mathbb{Z}$. This can be inferred from a construction by D.H. Lehmer [42].

$d\geq 4$ The above corollary implies that the solutions are contained in an exceptional subset of $\mathbb{P}^2$.

Of course the arguments given in cases $d=2,3$ are very ad hoc. It would be very interesting to have a more systematic treatment which the author is presently trying to work out.

For the proof of the following Corollary we refer to [80], Chapter 4.4.

4.4 Conjecture (Hall). *For any $\epsilon>0$ there exists $c(\epsilon)>0$ such that*

$$|x^3-y^2|>c(\epsilon)x^{\frac{1}{2}-\epsilon}$$

for any $x,y\in\mathbb{Z}$ with $y^2\neq x^3$.

It is known that $1/2$ is the best possible exponent.

The totally new and very remarkable ingredient in Vojta's conjecture is the occurrence of the term $h_K(P)$. It makes the following kind of statement possible.

4.5 Conjecture (Bombieri). *Let V be a projective variety over F and suppose that it is of general type. Then $V(F)$ is contained in a Zariski closed subset of V.*

Proof. Apply Vojta's conjecture to the case $D=0$, $A=K$ and $\epsilon=1/2$. Because V is of general type, K is pseudo ample. We find that for any $P\in V(F)$ outside a certain exceptional subvariety Z the height $h_K(P)$ is bounded. Because K is pseudo ample we conclude that $V(F)-Z$ is finite. □

Thus we see that in case $D=0$ Vojta's conjecture still makes highly non-trivial predictions thanks to the occurrence of $h_K(P)$. Roughly speaking the case $D=0$ can be seen as an example of diophantine approximation without actually approximating anything!

More particularly, when V is an algebraic curve of genus ≥ 2 defined over F, the number of F-rational points should be finite (Mordell's conjecture). This was proved by Faltings [21] in 1983, and later by Vojta in 1989 following an admirable adaptation of Siegel's diophantine approximation method. So it was Vojta himself who vindicated his insight that one can do diophantine approximation without approximants.

Finally there are some consequences of Vojta's conjecture which have recently been proved by Faltings as Theorems 5.5 and 5.6. We shall discuss them in the next section.

5 Results

In order to be able to appreciate the results we first establish a trivial result which is the analogue of Liouville's inequality. With the same notations as in Vojta's conjecture it reads as follows.

5.1 Theorem (Liouville). *Suppose that the divisor D is ample and defined over a finite extension L of F. Then we have for all $P \in V(F)$, $P \notin D$,*

$$\sum_v \lambda_{D,v}(P) < [L:F]h_D(P) + O(1).$$

Proof. Choose m such that mD is very ample. Let $f_0, f_1, \ldots, f_N$ be an L-basis of the linear system corresponding to mD. A good proximity function is given by

$$\lambda_{v,mD}(P) = \log \max_i \|f_i(P)\|_v.$$

Since the $\lambda_{v,mD}$ are functions bounded from below for all v and bounded from below by 0 for almost all v we find that

$$\sum_{v \in S} \lambda_{v,mD}(P) \leq \sum_v \log \max_i \|f_i(P)\|_v + O(1)$$

where for each v we have chosen an extension to L. The term on the right can now be bounded by $[L:F]h_{mD}(P) + O(1)$. Division by m on both sides yields our desired result. □

In the previous section we have already considered the Subspace theorem and its relation with Vojta's conjecture. In 1929 C.L. Siegel applied his theorem on the approximation of algebraic numbers by algebraic numbers from a fixed number field to the situation of algebraic curves. His result, as extended by Mahler, can be reinterpreted as follows.

5.2 Theorem (Siegel-Mahler). *Let C be a non-singular projective curve of genus ≥ 1 and defined over an algebraic number field F. Let S be a finite set of valuations on F. Let h be a height function on $C(\overline{F})$ and let D be a positive divisor on C defined over $\overline{F}$. Then, for any $\epsilon > 0$, we have*

$$\sum_{v \in S} \lambda_{D,v}(P) < \epsilon h(P)$$

for almost all $P \in C(F)$ (i.e. finitely many exceptions).

Proof. (sketch) First we embed C into its Jacobian variety J and let $\tilde{h}$ be a height on J. It is known that if we prove the theorem for $\tilde{h}$ restricted to C we have proved it for all h. Although Siegel had to work with a weaker result we can nowadays profit from Roth's theorem. It is a fairly direct consequence of Roth's theorem that

$$\sum_{v \in S} \lambda_{D,v}(P) < 3\mu\tilde{h}(P)$$

for almost all $P \in C(F)$, and where μ is the maximum multiplicity of the components of D. Actually we can have $2+\epsilon$ instead of the factor 3, but the latter will do. Let $\epsilon > 0$ and suppose that there exists an infinite subset $E \subset C(F)$ such that

$$\sum_{v \in S} \lambda_{D,v}(P) > \epsilon \tilde{h}(P) \tag{5.3}$$

for all $P \in E$. Choose $m \in \mathbb{N}$ such that $m^2\epsilon > 4\mu$. It is known via the weak Mordell-Weil theorem that $J(F)/mJ(F)$ is finite. Let $a_1, \ldots, a_n$ be a set of representatives of $J(F)/mJ(F)$. There exists a representative, a_1 say, such that $mQ + a_1 \in E$ for infinitely many $Q \in J(F)$. We denote this set of Q by E'. The covering $\omega : J \to J$ given by $\omega u = mu + a_1$ provides an unramified cover of C by an algebraic curve U. It follows from (5.3) that

$$\begin{aligned} \sum_{v \in S} \lambda_{D,v}(mQ + a_1) &> \epsilon \tilde{h}(mQ + a_1) \\ &= m^2 \epsilon \tilde{h}(Q) + O(1) \\ &> 4\mu \tilde{h}(Q) + O(1) \end{aligned} \tag{5.4}$$

for all $Q \in E'$. Notice that $\lambda_{D,v}(mQ + a_1)$ is a proximity function for $Q \in U(F)$ with respect to the divisor $\omega^* D$ on U. Because ω is unramified the maximum multiplicity of the components of $\omega^* D$ is also μ. Direct application of Roth's theorem to the curve U shows that

$$\sum_{v \in S} \lambda_{D,v}(mQ + a_1) > 3\mu \tilde{h}(Q)$$

has only finitely many solutions $Q \in U(F)$. This is in contradiction with (5.5). Hence (5.3) has only finitely many solutions, as asserted. □

It is well known that from Siegel's theorem we can derive the finiteness of the set of *integral points* on affine plane curves. This is done by taking for D the divisor at infinity. Consider for example the elliptic curve

$$y^2 = x^3 + Ax + B, \qquad A, B \in F$$

considered previously. In the section on Weil functions we saw that the function $\log \max(1, ||x||_v, ||y||_v)$ is a good proximity function to the point R at infinity, counted with multiplicity 3. Suppose we are only interested in points (x, y) on E with x, y in $\mathcal{O}_F$, the ring of integers of F. Take for S the set of infinite valuations. Then Siegel's theorem implies that

$$\sum_{v|\infty} \log \max(1, ||x||_v, ||y||_v) > \epsilon\, h(1 : x : y)$$

has finitely many solutions for any ϵ. If $x, y \in \mathcal{O}_F$ we see that the left hand side of the inequality equals $h(1 : x : y)$. Hence, if we take $\epsilon = 1/2$ say, we have a solution of the inequality. Siegel's theorem tells us that there are only finitely many such solutions.

In 1989 Vojta proved the Main Conjecture for the case of algebraic curves. The method used is an adaptation of a method of Siegel to the geometric case. Originally Vojta used some highly advanced results from the theory of arithmetic threefolds. However, Bombieri gave in 1990 a version which uses only fairly elementary notions of algebraic geometry. It seems that Vojta's ideas had broken a dam because Faltings very soon, in 1990 and 1991, produced two papers where the following two fascinating theorems are proved.

5.5 Theorem (Faltings). *Let A be an abelian variety over F and E a subvariety, also defined over F. Let h be a height on A and v a valuation on F. Let $\epsilon > 0$. Then we have*

$$\lambda_{E,v}(P) < \epsilon h(P)$$

for almost every point $P \in A(F) - E$.

This theorem settles a conjecture of S. Lang about integral points (with respect to E) on abelian varieties. In [80], Chapter 4.2 it is shown that Vojta's conjecture implies this theorem.

5.6 Theorem (Faltings). *Let A be an abelian variety defined over a number field F. Let X be a subvariety of A, also defined over F. Then the set $X(F)$ is contained in a finite union of translated abelian subvarieties of X.*

We observe that Theorem 5.6 is again a consequence of Vojta's conjecture. For this we remark that a subvariety of an abelian variety is either of general type or a translated abelian subvariety. Then we apply the following argument. If X is of general type, the exceptional subvariety is a lower dimensional subvariety. If a prime component X' of this exceptional subvariety is again of general type we repeat our argument for X'. When we have finally reached dimension zero, we are left with an exceptional subvariety which is a finite union of translated subtori. The only technical detail we have to be aware of is that X and other components of exceptional varieties may be singular. In that case we must resolve singularities to be able to apply Vojta's conjecture.

It is nice to notice that Theorem 5.6 partly solves two conjectures proposed by S. Lang. The first one is that an algebraic variety which is (Kobayashi) hyperbolic has only finite many F-rational points for a given number field F. In 1978 it was proved by M. Green that a subvariety X of an abelian variety which does not contain any translate of a positive dimensional abelian subvariety is indeed hyperbolic. Thus Theorem 5.6 settles Lang's conjecture for this particular type of hyperbolic variety. Theorem 5.6 is also interesting in another respect. Notice that $X(F) = A(F) \cap X$. We know that $A(F)$ is a finitely generated group. Thus the question of the nature of $X(F)$ can be considered as an example of problems where we intersect finitely generated subgroups of algebraic groups with their subvarieties. This is precisely the subject of Lang's second conjecture we alluded to. The conjecture, which has recently turned out to be a theorem, reads as follows.

5.7 Theorem. *Let G be either an abelian variety over $\mathbb{C}$ or a power of the multiplicative group $\mathbb{C}^*$. Let Γ be a subgroup of G of finite $\mathbb{Q}$-rank. Let V be a subvariety of G. Then $V \cap \Gamma$ is contained in a finite union of translates of algebraic subgroups of G.*

In the case when G is a power of the multiplicative group this was proved in 1984 by M. Laurent [41], using the Subspace theorem and a fair amount of Kummer theory. As for the case when G is an abelian variety defined over a number field it was shown in 1988 by M. Hindry [29] that the truth of the above theorem for the special case of finitely generated Γ implies the truth of the full theorem. The proof of this result is heavily based on work of M. Raynaud. Once that is done we only need to invoke Theorem 5.6 to prove the above theorem in case G is defined over a number field. The general case follows by a specialisation argument.

Chapter II

Diophantine Approximation and its Applications

by Rob Tijdeman

1 Upper Bounds for Approximations

Litt. [85], [58], [26] and [65].

As with many other areas, it is difficult to say when the development of the theory of diophantine approximation started. Diophantine equations have been solved long before Diophantos of Alexandria (perhaps A.D. 250) wrote his books on Arithmetics. Diophantos devised elegant methods for constructing one solution to an explicitly given equation, but he does not use inequalities. Archimedes's inequalities $3\frac{10}{71} < \pi < 3\frac{1}{7}$ and Tsu Ch'ung-Chih's (A.D. 430-501) estimate $\frac{355}{113} = 3.1415929\ldots$ for $\pi = 3.1415926\ldots$ are without any doubt early diophantine approximation results, but the theory of continued fractions does not have its roots in the construction methods for finding good rational approximations to π, but rather in the algorithm developed by Brahmagupta (A.D. 628) and others for finding iteratively the solutions of the Pell equation $x^2 - dy^2 = 1$. Euler proved in 1737 that the continued fraction expansion of any quadratic irrational number is periodic. The converse was proved by Lagrange in 1770. Lagrange deduced various inequalities on the convergents of irrational real numbers. In particular, he showed that every irrational real α admits infinitely many rationals p/q such that

$$\left|\alpha - \frac{p}{q}\right| < \frac{1}{q^2}. \tag{1.1}$$

Alternative proofs of this inequality are based on Farey sequences and on the Box principle. In 1842 Dirichlet used the latter principle to derive the following generalisation:

Suppose that α_{ij} $(1 \le i \le n,\ 1 \le j \le m)$ are nm real numbers and that $Q > 1$ is an integer. Then there exist integers $q_1, \ldots, q_m, p_1, \ldots, p_n$ with

$$1 \le \max(|q_1|, \ldots, |q_m|) < Q^{n/m}$$

and

$$|\alpha_{i1} q_1 + \cdots + \alpha_{im} q_m - p_i| \le Q^{-1} \qquad (1 \le i \le n).$$

(By taking $m = n = 1$ we find $1 \leq q < Q$, whence $|\alpha - p/q| < 1/q^2$.) A further generalisation resulted from the development of the geometry of numbers. Minkowski's Linear Forms Theorem (1896) reads as follows.

Suppose that (β_{ij}), $1 \leq i \leq n$, $1 \leq j \leq n$, is a real matrix with determinant ± 1. Let $A_1, \ldots, A_n$ be positive reals with $A_1 A_2 \cdots A_n = 1$. Then there exists an integer point $\mathbf{x} = (x_1, \ldots, x_n) \neq \mathbf{0}$ such that

$$|\beta_{i1}x_1 + \cdots + \beta_{in}x_n| < A_i \qquad (1 \leq i \leq n-1)$$

and

$$|\beta_{n1}x_1 + \cdots + \beta_{nn}x_n| \leq A_n.$$

(Dirichlet's Theorem corresponds with $n + m$ inequalities in $n + m$ variables.) The proof of Minkowski's Linear Forms Theorem is based on his Convex Body Theorem which says:

any convex compact set K in $\mathbb{R}^n$, symmetric about the origin, with the origin as interior point and with volume $V(K) \geq 2^n$, contains a point in $\mathbb{Z}^n \backslash \{\mathbf{0}\}$.

For $j = 1, 2, \ldots, n$, let $\lambda_j = \lambda_j(K)$ be the infimum of all $\lambda \geq 0$ such that λK contains j linearly independent integer points. The numbers $\lambda_1, \lambda_2, \ldots, \lambda_n$ are called the successive minima of K and satisfy $0 < \lambda_1 \leq \lambda_2 \leq \ldots \leq \lambda_n < \infty$. Minkowski's Convex Body Theorem implies that $\lambda_1^n V(K) \leq 2^n$. In 1907 Minkowski published the following refinement (Minkowski's Second Convex Body Theorem):

$$\frac{2^n}{n!} \leq \lambda_1 \lambda_2 \cdots \lambda_n V(K) \leq 2^n.$$

Both bounds can be attained.

The continued fraction algorithm provides a quick way to actually find rational approximations, but the theorems of Dirichlet and Minkowski don't. In [43] a polynomial-time basis reduction algorithm is introduced which is practical and theoretically still rather strong. It implies, as an effective version of Dirichlet's cited theorem:

There exists a polynomial-time algorithm that, given rational numbers α_{ij} $(1 \leq i \leq n,\ 1 \leq j \leq m)$ and Q satisfying $Q > 1$, finds integers $q_1, \ldots, q_m$, $p_1, \ldots, p_n$ for which

$$1 \leq \max(|q_1|, \ldots, |q_m|) < 2^{(m+n)^2/4m} Q^{n/m}$$

and

$$|\alpha_{i1}q_1 + \cdots + \alpha_{im}q_m - p_i| \leq Q^{-1} \qquad (1 \leq i \leq n).$$

The basis reduction algorithm has been used in the theory of factorisation of polynomials into irreducible factors, in optimisation theory and in several other areas. See [31]. In Section 3 some applications to diophantine equations will be mentioned.

2 Lower Bounds for Approximations

Litt. [8] and [65].

In 1844, Liouville proved that some number is transcendental. He derived this result from the following approximation theorem.

Suppose α is a real algebraic number of degree d. Then there is a constant $c(\alpha) > 0$ such that

$$\left|\alpha - \frac{p}{q}\right| > c(\alpha)q^{-d}$$

for every rational number p/q distinct from α.

(We tacitly assume $q > 0$). Liouville deduced that numbers like $\sum_{\nu=1}^{\infty} 2^{-\nu!}$ are transcendental. Liouville's Theorem implies that the inequality

$$\left|\alpha - \frac{p}{q}\right| < q^{-\mu} \tag{2.1}$$

has only finitely many rational solutions p/q if $\mu > d$. Thue showed in 1909 that (2.1) has only finitely many solutions if $\mu > d/2 + 1$. Then Siegel (1921) in his thesis showed that this is already true if $\mu > 2\sqrt{d}$. A slight improvement to $\mu > \sqrt{2d}$ was made by Dyson in 1947. Finally Roth proved in 1955 that (2.1) has only finitely many solutions if $\mu > 2$. If $d \geq 2$, then comparison with (1.1) shows that the number 2 is the best possible. If $d = 2$, then Liouville's Theorem is stronger than Roth's one. Chapter III of these notes contains a proof of Roth's Theorem.

In a series of papers published between 1965 and 1972, W.M. Schmidt proceeded an important step forward. One of his results is the following extension of Roth's Theorem.

Suppose α is a real algebraic number. Let $k \geq 1$ and $\delta > 0$. Then there are only finitely many algebraic numbers β of degree $\leq k$ with

$$|\alpha - \beta| < H(\beta)^{-(k+1+\delta)}$$

where $H(\beta)$ denotes the classical absolute height.

This result follows from the so-called Subspace Theorem which, in its simplest form, reads as follows.

Suppose $L_1(\mathbf{x}), \ldots, L_n(\mathbf{x})$ are linearly independent linear forms in $\mathbf{x} = (x_1, \ldots, x_n)$ with algebraic coefficients. Given $\delta > 0$, there are finitely many proper linear subspaces $T_1, \ldots, T_w$ of $\mathbb{R}^n$ such that every integer point $\mathbf{x} \neq \mathbf{0}$ with

$$|L_1(\mathbf{x}) \cdots L_n(\mathbf{x})| < |\mathbf{x}|^{-\delta}$$

lies in one of these subspaces.

($|\mathbf{x}|$ denotes the Euclidean length of $\mathbf{x}$.) In Chapter III it will be shown that Roth's Theorem is equivalent to the case $n = 2$ of the above stated Subspace Theorem.

Apart from Liouville's Theorem, all theorems stated in this section up to now are ineffective, that is, the method of proof does not enable us to determine the finitely many exceptions. However, the method makes it possible to derive upper bounds for the number of exceptions. I shall mention two results giving upper bounds for w in the Subspace Theorem.

The first result is due to Schmidt [66].

Let $L_1, \dots, L_n$ be linearly independent linear forms with coefficients in some algebraic number field of degree d. Consider the inequality

$$(2.2)\quad |L_1(\mathbf{x}) \cdots L_n(\mathbf{x})| < |\det(L_1, \dots, L_n)|\, |\mathbf{x}|^{-\delta} \quad \text{where } 0 < \delta < 1.$$

The set of solutions of (2.2) with

$$\mathbf{x} \in \mathbb{Z}^n, \qquad |\mathbf{x}| \gg \max((n!)^{8/\delta}, H(L_1), \dots, H(L_n))$$

is contained in the union of at most $[(2d)^{2^{26n}\delta^{-2}}]$ *proper linear subspaces of* $\mathbb{R}^n$.

The second result is due to Vojta [82]. Essentially it says that, apart from finitely many exceptions, which may depend on δ, the solutions of (2.2) are in the union of finitely many, at least in principle effectively computable, proper linear subspaces of $\mathbb{R}^n$ which are independent of δ.

Mahler derived in 1933 a p-adic version of the Thue-Siegel Theorem. Ridout and Schneider did so for Roth's Theorem. The p-adic versions of Schmidt's theorems have been proved by Schlickewei. We shall see that they have important applications. Vojta proved the p-adic assertion of the above mentioned result himself. For more information on the Subspace Theorem see Chapter IV of these notes.

Liouville's Theorem is also the starting point of a development of effective approximation methods. Hermite, in 1873, and Lindemann, in 1882, established the transcendence of the numbers e and π, respectively. The Theorem of Lindemann-Weierstrass (1885) says that $\beta_1 e^{\alpha_1} + \dots + \beta_n e^{\alpha_n} \neq 0$ for any distinct algebraic numbers $\alpha_1, \dots, \alpha_n$ and any nonzero algebraic numbers $\beta_1, \dots, \beta_n$.

A new development started in 1929 when Gelfond showed the transcendence of $2^{\sqrt{2}}$. In 1934 Gelfond and Schneider, independently of each other, proved the transcendence of α^β for α, β algebraic, $\alpha \neq 0, 1$ and β irrational. Alternatively, this result says that for any nonzero algebraic numbers $\alpha_1, \alpha_2, \beta_1, \beta_2$ with $\log \alpha_1, \log \alpha_2$ linearly independent over the rationals, we have

$$\beta_1 \log \alpha_1 + \beta_2 \log \alpha_2 \neq 0.$$

In 1966 Baker proved the transcendence of $e^{\beta_0} \alpha_1^{\beta_1} \cdots \alpha_n^{\beta_n}$ for any algebraic numbers $\alpha_1, \dots, \alpha_n$, other than 0 or 1, and $\beta_1, \dots, \beta_n$, provided that either $\beta_0 \neq 0$ or $1, \beta_1, \dots, \beta_n$ are linearly independent over the rationals. This follows from the theorem that

if $\alpha_1, \dots, \alpha_n$ are nonzero algebraic numbers such that their logarithms $\log \alpha_1, \dots, \log \alpha_n$ are linearly independent over the field of all rational numbers, then $1, \log \alpha_1, \dots, \log \alpha_n$ are linearly independent over the field of all algebraic numbers.

The above mentioned results have p-adic analogues and also analogues in the theory of elliptic functions. For example, Masser [46] showed that if a Weierstrass elliptic function $\wp(z)$ with algebraic invariants g_2 and g_3 and fundamental pair of periods ω_1, ω_2 has complex multiplication, then any numbers $u_1, \ldots, u_n$ for which $\wp(u_i)$ is algebraic for $i = 1, \ldots, n$ are either linearly dependent over $\mathbb{Q}(\omega_1/\omega_2)$ or linearly independent over the field of algebraic numbers.

The effective character of the results can be expressed in the form of transcendence measures. In 1972-77 Baker [7] derived the following important estimate.

Let $\alpha_1, \ldots, \alpha_n$ be nonzero algebraic numbers with degrees at most d and (classical absolute) heights at most $A_1, \ldots, A_n$ (all ≥ 2), respectively. Let $b_1, \ldots, b_n$ be rational integers of absolute values at most B (≥ 2). Put $\Lambda = b_1 \log \alpha_1 + \cdots + b_n \log \alpha_n$. Then either $\Lambda = 0$ or

$$\log |\Lambda| > -(16nd)^{200n} \left(\prod_{j=1}^{n} \log A_j \right) \log \left(\prod_{j=1}^{n-1} \log A_j \right) \log B.$$

The constants have been improved later on and the $\log(\prod \log)$ factor has been removed. The best bounds known at present, both in the complex and in the p-adic case, are due to Waldschmidt and his colleagues. They use a method going back to Schneider, whereas Baker's proof can be considered as an extension of Gelfond's proof of the transcendence of α^β. These linear form estimates have important applications to diophantine equations. Similar, but weaker, estimates have been obtained for linear forms in $u_1, \ldots, u_n$ such that u_i is a pole of $\wp(z)$ or $\wp(u_i)$ is an algebraic number, for $i = 1, \ldots, n$.

3 Applications to Diophantine Equations

Litt. [49], [72] and [67].

Results on diophantine approximations have been applied in various areas. I may refer to applications in algebraic number theory (class number problem, factorisation of polynomials), numerical mathematics (uniform distribution, numerical integration) and optimisation theory (when applying basis reduction algorithms, geometry of numbers). I shall deal here with applications to diophantine equations. This is very appropriate, since the last decade has also shown striking applications of arithmetic algebraic geometry to diophantine equations. It is quite likely that a merging of the theories of arithmetic algebraic geometry and diophantine approximations, as strived after in these Proceedings, would provide a new and solid basis for the theory of diophantine equations. This is a challenge for the young generation.

An immediate consequence of Thue's approximation result is as follows.

The equation

$$f(x, y) = a_0 x^n + a_1 x^{n-1} y + \cdots + a_n y^n = k \neq 0, \tag{3.1}$$

where $n \geq 3$, and $f(x, y) \in \mathbb{Q}[x, y]$ is irreducible, has only a finite number of solutions (in rational integers x, y).

The corresponding consequence of Roth's approximation result is that

the equation

$$f(x,y) = P(x,y), \tag{3.2}$$

where f is as above and $P(x,y) \in \mathbb{Q}[x,y]$ is any polynomial of degree $m < n-2$, has only finitely many solutions.

Schinzel used a suggestion of Davenport and Lewis to show that in the latter theorem the condition $m < n-2$ can be replaced by $m < n$. We note that, by a completely different method, Runge obtained a general result in case f is reducible, but $f - P$ is irreducible. Runge showed in 1887 that under these conditions there are only finitely many solutions provided that f is not a constant multiple of a power of an irreducible polynomial. Schinzel's result incorporates Runge's theorem and a result from Siegel's famous 1929-paper. It states that

if $f - P$ is irreducible, f is homogeneous of degree n and P is any polynomial of degree $< n$ such that (3.2) has infinitely many integer solutions, then f is a constant multiple of a power of a linear form or an irreducible quadratic form.

The p-adic analogues of approximation theorems led to important applications to S-unit equations. For simplicity I state results for rational integers, but corresponding results hold for integers from some algebraic number field or even finitely generated integral domains, see [20]. Let $p_1, \dots, p_s$ be given prime numbers and denote by S_0 all rational integers composed of $p_1, \dots, p_s$. In 1933 Mahler showed that

the equation

$$x + y = z \quad \text{in} \quad x, y, z \in S_0 \quad \text{with} \quad \gcd(x,y) = 1 \tag{3.3}$$

has only a finite number of solutions.

He applied his p-adic version of the Thue-Siegel method. A general and in some respect best possible result was proved by Evertse in 1984. He used Schlickewei's p-adic version of Schmidt's Subspace Theorem to show that

for any reals c, d with $c > 0$, $0 \leq d < 1$ and any positive integer n, there are only finitely many $(x_1, \dots, x_n) \in \mathbb{Z}^n$ such that (i) $x_1 + \dots + x_n = 0$, (ii) $x_{i_1} + \dots + x_{i_t} \neq 0$ *for each proper non-empty subset* $\{i_1, \dots, i_t\}$ *of* $\{1, \dots, n\}$, (iii) $\gcd(x_1, \dots, x_n) = 1$ *and* (iv)

$$\prod_{k=1}^{n} \Big(|x_k| \prod_{p \in S} |x_k|_p\Big) \leq c \max_{1 \leq k \leq n} |x_k|^d. \tag{3.4}$$

For the many diverse applications of this and related results I refer the reader to the survey paper [20].

The proofs of the above mentioned results are ineffective. So it is impossible to derive upper bounds for the size of the solutions by following the proofs. However, it is possible to give upper bounds for the numbers of solutions. A remarkable feature is that these bounds depend on very few parameters. Lewis and Mahler derived in 1961 an explicit upper bound for the number of solutions of (3.3) depending on $p_1, \ldots, p_s$. In 1984, Evertse improved this upper bound to $3 \times 7^{2s+3}$ which depends only on s. Schlickewei [63] gave an upper bound for the number of solutions of (3.4) with $c = 1$, $d = 0$ in Evertse's theorem. This bound depends only on n and s.

Bombieri and Schmidt [10] proved that the number of primitive solutions of the Thue equation (3.1) is bounded by $c_1 n^{s+1}$ where c_1 is an explicitly given absolute constant and s denotes the number of distinct prime factors of k. Recently, Schmidt derived good upper bounds for the number of integer points on elliptic curves.

Effective methods make it possible to compute upper bounds for the solutions themselves. In 1960, Cassels obtained an effective result on certain special cases of equation (3.3) by applying Gelfond's result on the transcendence of α^β. Baker's estimates in 1967 on linear forms of logarithms in algebraic numbers caused a breakthrough. Baker himself gave upper bounds for the solutions of the Thue equation (3.1) and the super-elliptic equation

$$y^m = P(x) \tag{3.5}$$

where $m \geq 2$ is a fixed given positive integer and $P(x) \in \mathbb{Z}[x]$ a polynomial with at least three simple roots if $m = 2$, and at least two simple roots if $m \geq 3$. (Later, the conditions were weakened.) Baker and Coates derived an upper bound for the number of integer points on some curve of genus 1. (This bound has been sharpened by Schmidt.) Coates used the p-adic estimates to obtain bounds for the Thue-Mahler equation (3.1) with m unknown and in S, equation (3.3), and the equations $y^2 = x^3 + k$ with k unknown and in S. There are many later generalisations and improvements of the bounds.

Baker's sharpening made it possible to deal with diophantine equations which cannot be treated by the mentioned ineffective method. E.g. Schinzel and Tijdeman showed that equation (3.5) with m, x, y variables and $P(x) \in \mathbb{Z}[x]$ a given polynomial with at least two distinct roots implies that m is bounded. Tijdeman also showed that the Catalan equation $x^m - y^n = 1$ in integers m, n, x, y all > 1 implies that x^m is bounded by some effectively computable number. The bounds obtained in equations involving a power with both base and exponent variable are, however, so large that it is not yet possible to solve such equations in practice.

The best bounds for linear forms known at present make it, however, possible to solve Mahler's equation (3.3) and Thue and Thue-Mahler equations (3.1) completely. Additional algorithms are needed to achieve this. To give a typical example, Tzanakis and de Weger [77] considered the Weierstrass equation $y^2 = x^3 - 4x + 1$. They reduced it to some Thue equations, of which $f(x,y) = x^4 - 12x^2y^2 - 8xy^3 + 4y^4 = 1$ is a typical example. This leads to some equations

$$x - \vartheta y = \epsilon_1^{a_1} \epsilon_2^{a_2} \epsilon_3^{a_3},$$

where ϵ_1, ϵ_2, ϵ_3 is a fixed fundamental set of units of $\mathbb{Q}(\vartheta)$ and ϑ is a zero of $f(x,1)$. A suitable linear form estimate yields $\max(|a_1|, |a_2|, |a_3|) < 10^{41}$. Subsequently the basis reduction algorithm of Lenstra, Lenstra and Lovász is applied. The first time yields an upper bound 72, the second time an upper bound 10. Checking the remaining values yields four solutions, $(0,1)$, $(1,-1)$, $(3,1)$ and $(-1,3)$. They correspond to the solutions $(x, \pm y) = (2,1)$, $(10,31)$, $(1274,45473)$, $(114,1217)$, respectively, of the equation $y^2 = x^3 - 4x + 1$. In this way Tzanakis and de Weger determined the 22 integral solutions of the equation. For an introduction to the available additional techniques, I refer to [86].

The following results illustrate the power of the described method.

(a) [86] The Mahler equation $x + y = z$ subject to $\gcd(x, y) = 1$, $x \leq y$, xyz composed of the primes 2, 3, 5, 7, 11 and 13, has exactly 545 solutions of which all large ones are explicitly stated.

(b) [86] The equation $x + y = z^2$ in integers $x \geq y$, $z > 0$ such that both x and y are composed of primes 2, 3, 5 and 7 and that $\gcd(x, y)$ is squarefree, has exactly 388 solutions. The largest one is $(x, y, z) = (199290375, -686, 14117)$.

(c) [78] The Thue-Mahler equation

$$x^3 - 23x^2y + 5xy^2 + 24y^3 = \pm 2^{z_1}3^{z_2}5^{z_3}7^{z_4}$$

in integers x, y, z_1, z_2, z_3 and z_4 with $x \geq 0$ has exactly 72 solutions. The largest is given by $(x, y) = (48632, -3729)$.

Chapter III

Roth's Theorem

by Rob Tijdeman

This chapter is based on Schmidt [65] and as to the proof of Lemma 1.6 on Wirsing [89].

1 The Proof

1.1 Theorem (Roth, [61]). *Suppose α is real and algebraic of degree $d \geq 2$. Then for each $\delta > 0$, the inequality*

$$\left|\alpha - \frac{p}{q}\right| < q^{-(2+\delta)} \tag{1.2}$$

has only finitely many solutions in rationals p/q.

1.3 Remarks.

(1) By Dirichlet's Theorem the number 2 in (1.2) is best possible.

(2) If α is of degree 2, then Liouville's Theorem implies the stronger inequality

$$\left|\alpha - \frac{p}{q}\right| > c(\alpha)q^{-2} > 0 \tag{1.4}$$

for all rationals p/q. For no single α of degree ≥ 3 do we know whether (1.4) holds. It is likely that (1.4) is false for every such α (that is, that every such α has unbounded partial quotients).

(3) Lang conjectured in 1965 that for α of degree ≥ 3

$$\left|\alpha - \frac{p}{q}\right| < q^{-2}(\log q)^{-K}$$

has only finitely many solutions if $K > 1$, or at least if $K > K_0(\alpha)$. □

The first lemma is a straightforward application of the box principle. For a matrix $A = (a_{jk})$ with rational integer coefficients, put $|A| = \max |a_{jk}|$. For an integer vector $\mathbb{Z} = (z_1, \ldots, z_N)$, put $|\mathbf{z}| = \max(|z_1|, \ldots, |z_N|)$.

1.5 Lemma (Siegel). *Let A be an $M \times N$ matrix with rational integer coefficients, not all zero, and suppose that $N > M$. Then there is a $\mathbf{z} \in \mathbb{Z}^N$ with*

$$A\mathbf{z} = 0, \qquad \mathbf{z} \neq 0, \qquad |\mathbf{z}| \leq (N|A|)^{M/(N-M)}.$$

Proof. Put $Z = [(N|A|)^{M/(N-M)}]$ and $L_j(\mathbf{z}) = \sum_{k=1}^{N} a_{jk} z_k$ for $j = 1, \dots, m$ and $\mathbf{z} = (z_1, \dots, z_N)$. For $\mathbf{z} \in \mathbb{Z}^N$ with $0 \leq z_k \leq Z$ $(1 \leq k \leq N)$ there are at most $N|A|Z + 1$ possible values for $L_j(\mathbf{z})$. Hence $A\mathbf{z}$ takes at most $(N|A|Z+1)^M$ values. Since $(N|A|Z+1)^M \leq (N|A|)^M (Z+1)^M < (Z+1)^N$, there are $\mathbf{z}^{(1)} \neq \mathbf{z}^{(2)}$ in the considered set with $A\mathbf{z}^{(1)} = A\mathbf{z}^{(2)}$. Then $\mathbf{z} := \mathbf{z}^{(1)} - \mathbf{z}^{(2)}$ satisfies the conditions of the lemma. □

The second lemma states that most of the values $i_1/d_1 + \dots + i_m/d_m$ with $0 \leq i_h \leq d_h$ $(h = 1, \dots, m)$ are close to $m/2$.

1.6 Lemma (Combinatorial Lemma). *Suppose $d_1, \dots, d_m \in \mathbb{Z}_{\geq 1}$ and $0 < \epsilon < 1$. Then the number of tuples $(i_1, \dots, i_m) \in \mathbb{Z}^m$ with*

$$0 \leq i_h \leq d_h \quad (h = 1, \dots, m) \qquad \text{and} \qquad \left| \sum_{h=1}^{m} \frac{i_h}{d_h} - \frac{m}{2} \right| \geq \epsilon m$$

is at most $(d_1+1)\cdots(d_m+1)/(4m\epsilon^2)$.

Proof. We may consider $i_1, \dots, i_m$ as independent stochastic variables such that i_h is uniformly distributed on $\{0, \dots, d_h\}$. Define the stochastic variable $X = \sum_{h=1}^{m} i_h/d_h$. Then X has expectation $\mu X = m/2$ and variance

$$\sigma^2 = \mathrm{Var}(i_1/d_1) + \dots + \mathrm{Var}(i_m/d_m).$$

We have

$$\mathrm{Var}(i_h/d_h) = \sum_{i_h=0}^{d_h} \left(\frac{i_h}{d_h} - \frac{1}{2} \right)^2 \frac{1}{d_h+1} = \frac{2d_h+1}{6d_h} - \frac{1}{4} \leq \frac{1}{4}.$$

Hence $\sigma^2 \leq m/4$. By Kolmogorov's generalisation of Chebyshev's inequality, we have $\mathrm{Prob}(|X - \mu| \geq c) \leq \sigma^2/c^2$. Thus

$$\mathrm{Prob}(|X - m/2| \geq \epsilon m) \leq \frac{1}{4m\epsilon^2}.$$

□

For a polynomial $P(\mathbf{X}) = P(X_1, \dots, X_m) \in \mathbb{Z}[X_1, \dots, X_m]$ and $\mathbf{i} = (i_1, \dots, i_m) \in \mathbb{Z}^m_{\geq 0}$, put

$$\begin{aligned} P_{\mathbf{i}}(\mathbf{X}) &= \frac{1}{i_1! \cdots i_m!} \frac{\partial^{i_1}}{\partial X_1^{i_1}} \cdots \frac{\partial^{i_m}}{\partial X_m^{i_m}} P(\mathbf{X}) \\ &= \sum_{j_1=0}^{d_1} \cdots \sum_{j_m=0}^{d_m} \binom{j_1}{i_1} \cdots \binom{j_m}{i_m} C(j_1, \dots, j_m) X_1^{j_1-i_1} \cdots X_m^{j_m-i_m}. \end{aligned}$$

It follows that $P_{\mathbf{i}}(\mathbf{X})$ has integer coefficients. Let $\alpha_1, \dots, \alpha_m \in \mathbb{C}$ and $d_1, \dots, d_m \in \mathbb{Z}_{\geq 1}$. The *index* $i(P)$ of P with respect to $\underline{\alpha} = (\alpha_1, \dots, \alpha_m)$ and $(d_1, \dots, d_m)$ is the least value σ for which there is a tuple $\mathbf{i} = (i_1, \ldots, i_m)$ with $i_1/d_1 + \cdots + i_m/d_m = \sigma$ and $P_{\mathbf{i}}(\underline{\alpha}) \neq 0$. (If $P = 0$, then the index is defined to be ∞.) Note that $i(PQ) = i(P) + i(Q)$ and $i(P+Q) \geq \min(i(P), i(Q))$.

The third lemma provides the construction of a polynomial with high index at some given point. For $P \in \mathbb{Z}[X_1, \ldots, X_m]$ we denote the maximum of the absolute values of the coefficients of P by $|P|$.

1.7 Lemma (Index Theorem). *Suppose α is an algebraic integer of degree $d \geq 2$. Let $\epsilon > 0$, and let $m \in \mathbb{Z}$ with $m \geq d/2\epsilon^2$. Let $d_1, \ldots, d_m \in \mathbb{N}$. Then there is a polynomial P in $\mathbb{Z}[X_1, \ldots, X_m]$, $P \neq 0$, such that*

(i) *P has degree $\leq d_h$ in X_h,*

(ii) *P has index $> m(1-\epsilon)/2$ with respect to $(\alpha, \ldots, \alpha)$ and $(d_1, \ldots, d_m)$,*

(iii) *$|P| \leq C_1^{d_1 + \cdots + d_m}$.*

Proof. Write $P(X_1, \ldots, X_m) = \sum_{j_1=0}^{d_1} \cdots \sum_{j_m=0}^{d_m} z(j_1, \ldots, j_m) X_1^{j_1} \cdots X_m^{j_m}$, where the $z(j_1, \ldots, j_m)$ are integers which have to be determined such that (ii) holds, i.e., $P_{\mathbf{i}}(\underline{\alpha}) = 0$ for $i_1/d_1 + \cdots + i_m/d_m \leq m(1-\epsilon)/2$. By taking all these expressions together, we obtain

$$A_0 \mathbf{z} + \alpha A_1 \mathbf{z} + \cdots + \alpha^{d_1 + \cdots + d_m} A_{d_1 + \cdots + d_m} \mathbf{z} = \mathbf{0}$$

where the A_i are $M \times N$ integer matrices with $|A_i| \leq 4^{d_1 + \cdots + d_m}$, where $N = (d_1 + 1) \cdots (d_m + 1)$ and M is the number of tuples $\mathbf{i}$ with $i_1/d_1 + \cdots + i_m/d_m \leq m(1-\epsilon)/2$. Using that α is an algebraic number of degree d, we get

$$B_0 \mathbf{z} + \alpha B_1 \mathbf{z} + \cdots + \alpha^{d-1} B_{d-1} \mathbf{z} = \mathbf{0}$$

where the B_i are integer $M \times N$ matrices with $|B_i| \leq C_2^{d_1 + \cdots + d_m}$. Since $1, \alpha, \ldots, \alpha^{d-1}$ are $\mathbb{Z}$-linearly independent, we have $B_0 \mathbf{z} = \mathbf{0}$, $B_1 \mathbf{z} = \mathbf{0}, \ldots, B_{d-1} \mathbf{z} = \mathbf{0}$. Hence $B\mathbf{z} = \mathbf{0}$ where B is a $dM \times N$ integer matrix with $|B| \leq C_2^{d_1 + \cdots + d_m}$. By the Combinatorial Lemma, we have

$$M \leq \frac{(d_1+1) \cdots (d_m+1)}{4m\epsilon^2} = \frac{N}{4m\epsilon^2} \leq \frac{N}{2d}.$$

Now Siegel's lemma implies that there is a non-zero integer vector $\mathbf{z}$ such that

$$B\mathbf{z} = \mathbf{0}, \qquad |\mathbf{z}| \leq (N|B|)^{dM/(N-dM)} \leq N|B| \leq C_3^{d_1 + \cdots + d_m}.$$

Note that the constants C_1, C_2 and C_3 depend only on α. □

The fourth lemma gives a sufficient condition for a polynomial to have a small index with respect to the approximation vector $(p_1/q_1, \ldots, p_m/q_m)$ and $(d_1, \ldots, d_m)$.

1.8 Lemma (Roth). *Let $m \in \mathbb{Z}$, $m \geq 1$, and $\epsilon > 0$. There exists a number $C_4 = C_4(m, \epsilon) > 1$ with the following property: Let $d_1, \ldots, d_m$ be positive integers with $d_h \geq C_4 d_{h+1}$ for $h = 1, \ldots, m-1$. Let $(p_1, q_1), \ldots, (p_m, q_m)$ be pairs of coprime integers with*

(1.9) $$q_h^{d_h} \geq q_1^{d_1} \quad \text{and} \quad q_h \geq 2^{2mC_4} \quad \text{for} \quad h = 1, \ldots, m.$$

Let $P(X_1, \ldots, X_m) \in \mathbb{Z}[X_1, \ldots, X_m]$ be a polynomial of degree $\leq d_h$ in X_h for $h = 1, \ldots, m$ and with

(1.10) $$|P|^{C_4} \leq q_1^{d_1}, \qquad P \neq 0.$$

Then the index of P with respect to $(p_1/q_1, \ldots, p_m/q_m)$ and $(d_1, \ldots, d_m)$ is $\leq \epsilon$.

Proof. We use induction on m. The case $m = 1$. Write $P(X) = (q_1X - p_1)^l R(X)$ where l/d_1 is the index of P with respect to p_1/q_1 and d_1. The polynomial $R(X)$ has integer coefficients by Gauss's Lemma. Since q_1^l divides the leading coefficient of P, we obtain $q_1^l \leq |P| \leq q_1^{d_1/C_4}$ by (1.10). By choosing $C_4(1,\epsilon) = \epsilon^{-1}$ we get $l/d_1 \leq \epsilon$.

The induction step. Let $m \geq 2$. Suppose Roth's Lemma is true for $m-1$, but not for m, and $C_4(m-1,\delta)$ has been defined for $0 < \delta < 1$. We shall apply the case $m-1$ of Roth's Lemma to some Wronskian which is constructed as follows. Consider a decomposition

$$P(X_1,\dots,X_m) = \sum_{j=1}^{k} \phi_j(X_1,\dots,X_{m-1})\psi_j(X_m) \tag{1.11}$$

where $\phi_1,\dots,\phi_k$ and $\psi_1,\dots,\psi_k$ are polynomials with rational coefficients and k is minimal. Since the choice $\psi_j = X_m^{j-1}$ $(j = 1,\dots,d_m+1)$ is possible, we have

$$k \leq d_m + 1. \tag{1.12}$$

The minimality of k implies that both $\phi_1,\dots,\phi_k$ and $\psi_1,\dots,\psi_k$ are linearly independent over the reals. We consider differential operators

$$\Delta = \frac{\partial^{i_1+\dots+i_m}}{\partial X_1^{i_1}\cdots\partial X_m^{i_m}}$$

and call $i_1 + \dots + i_m$ its order. A *(generalised) Wronskian* of $\phi_1,\cdots,\phi_k$ is any determinant of the form

$$\det(\Delta_i\phi_j) \qquad (1 \leq i \leq k,\ 1 \leq j \leq k)$$

where $\Delta_1,\dots,\Delta_k$ are operators as above, with Δ_i of order $\leq i-1$ for $i = 1,\dots,k$. We shall choose the Δ_i in such a way that $\det(\Delta_i\phi_j)$ does not vanish. Such a choice is possible in view of the following lemma.

1.13 Lemma. *Suppose that $\phi_1,\dots,\phi_k$ are rational functions in $X_1,\dots,X_m$ with real coefficients, and linearly independent over the reals. Then at least one Wronskian of $\phi_1,\dots,\phi_k$ is not identically zero.*

Proof. We use induction on k. If $k = 1$, then $\det(\Delta_i\phi_j) = \phi_1 \neq 0$. Suppose now that $\phi_1,\dots,\phi_k$ are $k \geq 2$ rational functions satisfying the hypotheses of Lemma 1.13. Then 1, $\phi_2/\phi_1,\dots,\phi_k/\phi_1$ are also linearly independent over the reals. It follows that

$$\frac{\partial}{\partial X_j}\frac{\phi_2}{\phi_1},\dots,\frac{\partial}{\partial X_j}\frac{\phi_k}{\phi_1}$$

are linearly independent over the reals for some j with $1 \leq j \leq m$. By the induction hypothesis there exists a Wronskian of these functions which is not identically zero. This induces a Wronskian of 1, $\phi_2/\phi_1,\dots,\phi_k/\phi_1$ which is not identically zero. It follows from a simple induction argument that for any rational function ϕ a Wronskian of $\phi\phi_1,\phi\phi_2,\dots,\phi\phi_k$ can be written as a linear combination of Wronskians of $\phi_1,\dots,\phi_k$ with coefficients which are rational functions involving only the partial derivatives of ϕ. By taking $\phi = 1/\phi_1$ we can conclude that there exists a (generalised) Wronskian of ϕ_1, $\phi_2,\dots,\phi_k$ which is not identically zero. □

We can now define the Wronskian to which Roth's Lemma for $m-1$ will be applied. By Lemma 1.13 there exist operators

$$\Delta_i' = \frac{1}{i_1! \cdots i_{m-1}!} \frac{\partial^{i_1+\cdots+i_{m-1}}}{\partial X_1^{i_1} \cdots \partial X_{m-1}^{i_{m-1}}}$$

with orders $i_1 + \cdots + i_{m-1} \leq i-1 \leq k-1$ for $i = 1, \ldots, k$, such that

$$U(X_1, \ldots, X_{m-1}) := \det(\Delta_i' \phi_j)_{1 \leq i \leq k,\ 1 \leq j \leq k} \neq 0.$$

By Lemma 1.13

$$V(X_m) := \det\left(\frac{1}{(i-1)!} \frac{\partial^{i-1}}{\partial X_m^{i-1}} \psi_j(X_m)\right)_{1 \leq i \leq k,\ 1 \leq j \leq k} \neq 0,$$

since all the other generalised Wronskians of the prescribed type vanish. Put

$$W(X_1, \ldots, X_m) = \det\left(\frac{1}{(j-1)!} \frac{\partial^{j-1}}{\partial X_m^{j-1}} \Delta_i' P\right)_{1 \leq i \leq k,\ 1 \leq j \leq k}$$

Then, by (1.11),

$$W(X_1, \ldots, X_m) = U(X_1, \ldots, X_{m-1}) V(X_m) \neq 0.$$

The entries in the determinant defining W are of the type $P_{i_1 \cdots i_{m-1} j-1}$, whence these entries have rational integer coefficients. Thus W is a polynomial with rational integer coefficients.

The determinant W is a polynomial in P and its partial derivatives. Hence a lower bound for the index ϑ of P with respect to $(p_1/q_1, \ldots, p_m/q_m)$ and $(d_1, \ldots, d_m)$ implies a lower bound for the index Θ of W with respect to the same values. We shall compute such a lower bound. On using the conditions of Lemma 1.8 we obtain, for $i_1 + \cdots + i_{m-1} \leq k-1 \leq d_m$, that

$$\begin{aligned} i(P_{i_1 \cdots i_{m-1} j-1}) &\geq \vartheta - \frac{i_1}{d_1} - \frac{i_2}{d_2} - \cdots - \frac{i_{m-1}}{d_{m-1}} - \frac{j-1}{d_m} \\ &\geq \vartheta - \frac{i_1 + \ldots + i_{m-1}}{d_{m-1}} - \frac{j-1}{d_m} \geq \vartheta - \frac{d_m}{d_{m-1}} - \frac{j-1}{d_m} \geq \vartheta - C_4^{-1} - \frac{j-1}{d_m}. \end{aligned}$$

Note that this index is also non-negative. By expanding W we get using the formulas for the index of the sum and of the product of functions,

$$\begin{aligned} &(1.14) \qquad \Theta \geq \sum_{j=0}^{k-1} \max(\vartheta - C_4^{-1} - \frac{j}{d_m}, 0) \geq -kC_4^{-1} + \sum_{j=0}^{[\vartheta k]} (\vartheta - \frac{j}{d_m}) \\ &(1.15) \qquad \geq -kC_4^{-1} + \frac{\vartheta}{2}([\vartheta k] + 1) \geq -kC_4^{-1} + \frac{1}{2}\vartheta^2 k. \end{aligned}$$

Next we shall use the induction hypothesis to show that Θ cannot be large. Let $0 < \delta < \vartheta < 1$. The factorisation $W = UV$ induces the factorisation

$$W(X_1, \ldots, X_m) = U^*(X_1, \ldots, X_{m-1}) V^*(X_m)$$

where U^* and V^* have rational integer coefficients. Note that

$$|P_{i_1\cdots i_{m-1}j-1}| \le 2^{d_1+\cdots+d_m}|P| \le 2^{d_1+\cdots+d_m} q_1^{d_m/C_4},$$

by (1.10). Since the number of summands in the determinant expansion of W does not exceed $k! \le k^{k-1} \le k^{d_m} \le 2^{kd_m}$, it follows that

$$|W| \le 2^{kd_m}(2^{d_1+\cdots+d_m} q_1^{d_1/C_4})^k \le (2^{2m} q_1^{1/C_4})^{d_1 k} < q_1^{2d_1 k/C_4}$$

in view of (1.9). This yields the estimates

$$|U^*| \le q_1^{2d_1 k/C_4}, \qquad |V^*| \le q_1^{2d_1 k/C_4} \le q_m^{2d_m k/C_4}, \qquad \text{by (1.9).}$$

We will now apply the induction hypothesis to $U^*(X_1, \ldots, X_{m-1})$ with respect to $(p_1/q_1, \ldots, p_{m-1}/q_{m-1})$ and $(kd_1, \ldots, kd_{m-1})$. If we choose $C_4(m, \epsilon)$ in such a way that $C_4(m,\epsilon) \ge 2C_4(m-1, \delta)$, then

$$|U^*| \le q_1^{2d_1 k/C_4(m,\epsilon)} \le q_1^{d_1 k/C_4(m-1,\delta)}.$$

We conclude therefore that the index of U^* with respect to $(p_1/q_1, \ldots, p_{m-1}/q_{m-1})$ and $(d_1 k, \ldots, d_{m-1}k)$ is at most δ. Similarly we can conclude by applying the induction hypothesis for $m = 1$ to $V(X_m)$ that the index of V^* with respect to p_m/q_m and $d_m k$ is at most δ. Since $W = U^*V^*$, the index of W with respect to $(p_1/q_1, \ldots, p_m/q_m)$ and $(d_1 k, \ldots, d_m k)$ is at most 2δ. Hence

$$\Theta \le 2\delta k, \tag{1.16}$$

where, as before, Θ is the index of W with respect to $(p_1/q_1, \ldots, p_m/q_m)$ and $(d_1, \ldots, d_m)$. A suitable choice of δ yields a contradiction to our assumption $\vartheta > \epsilon$. Namely, choose $C_4(m, \epsilon)$ so large that $C_4(m, \epsilon) > 4\epsilon^{-2}$. Then, by (1.14),

$$\Theta \ge -kC_4^{-1} + \epsilon^2 k/2 \ge \epsilon^2 k/4.$$

On the other hand, choosing $\delta < \epsilon^2/8$, we obtain from (1.16) that

$$\Theta < \epsilon^2 k/4.$$

□

Suppose α is an algebraic number of degree $d \ge 2$ such that, for some δ with $0 < \delta < 1$,

$$\left|\alpha - \frac{p}{q}\right| < q^{-2-\delta} \tag{1.17}$$

has infinitely many rational solutions p/q. We shall use the Index Theorem to construct a polynomial P with very high index with respect to $(\alpha, \ldots, \alpha)$ and arbitrary $(d_1, \ldots, d_m)$. We shall show below that this implies that for a suitable choice of solutions $p_1/q_1, \ldots, p_m/q_m$ of (1.17) the polynomial P has still high index with respect to $(p_1/q_1, \ldots, p_m/q_m)$ and $(d_1, \ldots, d_m)$. On the other hand, we shall choose $d_1, \ldots, d_m$ in such a way that P has to have low index with respect to $(p_1/q_1, \ldots, p_m/q_m)$ and $(d_1, \ldots, d_m)$ by Roth's Lemma. This contradiction will complete the proof.

Proof. (of Roth's Theorem) Let α be an algebraic number with denominator a_0 satisfying (1.2). Then

$$\left|(a_0\alpha) - \frac{a_0 p}{q}\right| < \frac{a_0}{q^{2+\delta}} < \frac{1}{q^{2+\delta/2}}$$

for $q \geq q_0(\alpha)$. Hence we conclude that it suffices to prove Roth's Theorem for algebraic integers α. We assume that α is an algebraic integer of degree $d \geq 2$ with $|\alpha| < 1$ and δ some number with $0 < \delta < 1/2$ such that (1.17) has infinitely many rational solutions p/q.

Let P be the polynomial constructed in the Index Theorem with respect to α, $\epsilon = \delta/12$, $m > d/2\epsilon^2$ and arbitrary $d_1, \ldots, d_m$. Then P has index $> m(1-\epsilon)/2$ with respect to $(\alpha, \ldots, \alpha)$ and $(d_1, \ldots, d_m)$. We first choose solutions $p_1/q_1, \ldots, p_m/q_m$ of (1.17) and integers $d_1, \ldots, d_m$ as follows:

(a) choose (p_1, q_1) with

$$q_1 > \max((6C_1)^{1/\epsilon}, C_1^m, 2^{2mC_4}) \tag{1.18}$$

(b) choose solutions $(p_2, q_2), \ldots, (p_m, q_m)$ such that

$$\log q_{h+1} > (1+\epsilon)C_4 \log q_h \qquad (1 \leq h \leq m-1) \tag{1.19}$$

[This step makes the proof ineffective.]

(c) choose d_1 so large that

$$\epsilon d_1 \log q_1 \geq \log q_m \tag{1.20}$$

(d) for $h = 2, \ldots, m$, choose d_h such that

$$q_1^{d_1} \leq q_h^{d_h} < q_1^{d_1(1+\epsilon)}. \tag{1.21}$$

(This is possible since $q_1^{\epsilon d_1} \geq q_m \geq q_h$.)

The conditions of Roth's Lemma are satisfied, since (1.18) and (1.21) imply

$$\frac{d_h}{d_{h+1}} = \frac{d_h \log q_h}{d_{h+1} \log q_{h+1}} \cdot \frac{\log q_{h+1}}{\log q_h} \geq \frac{1}{1+\epsilon} \cdot (1+\epsilon)C_4 = C_4 \geq 1,$$

whence $d_1 \geq d_2 \geq \cdots \geq d_m$, and further (1.21) and (1.18) imply

$$q_h^{d_h} \geq q_1^{d_1} > 2^{2mC_4 d_1} > 2^{2mC_4 d_h}.$$

Hence the polynomial P has index $\leq \epsilon$ with respect to $(p_1/q_1, \ldots, p_m/q_m)$ and to $(d_1, \ldots, d_m)$.

We finally show, in order to obtain a contradiction, that P has index $> \epsilon$ with respect to $(p_1/q_1, \ldots, p_m/q_m)$ and $(d_1, \ldots, d_m)$. So we have to prove that for $\mathbf{i}$ with

$$\frac{i_1}{d_1} + \cdots + \frac{i_m}{d_m} \leq \epsilon \tag{1.22}$$

we have $P_{\mathbf{i}}(p_1/q_1, \ldots, p_m/q_m) = 0$. Note that

$$P_{\mathbf{i}}(\underline{\alpha}) = \sum_{\mathbf{j}} P_{\mathbf{j}}(\mathbf{0}) \binom{j_1}{i_1} \cdots \binom{j_m}{i_m} \alpha^{j_1 - i_1} \cdots \alpha^{j_m - i_m},$$

whence, using $|\alpha| < 1$,

$$(1.23) \qquad |P_{\mathbf{i}}(\underline{\alpha})| \le |P| \max_{\mathbf{j}} \binom{j_1}{i_1} \cdots \binom{j_m}{i_m} \le (2C_1)^{d_1+\cdots+d_m} \le (2C_1)^{md_1}$$

where the maximum extends over all $\mathbf{j}$ with $j_h \le d_h$ for $h = 1, \ldots, m$. Expand $P_{\mathbf{i}}(\mathbf{X})$ in a Taylor series around $\underline{\alpha} = (\alpha, \ldots, \alpha)$,

$$P_{\mathbf{i}}(\mathbf{X}) = \sum_{j} P_{\mathbf{j}}(\underline{\alpha}) \binom{j_1}{i_1} \cdots \binom{j_m}{i_m} (X_1 - \alpha)^{j_1 - i_1} \cdots (X_m - \alpha)^{j_m - i_m}.$$

According to the Index Theorem, $P_{\mathbf{j}}(\underline{\alpha}) = 0$, if $j_1/d_1 + \cdots + j_m/d_m \le m(1-\epsilon)/2$, so certainly if $(j_1 - i_1)/d_1 + \cdots + (j_m - i_m)/d_m \le m(1 - 3\epsilon)/2$. Furthermore, by (1.23),

$$\sum_{\mathbf{j}} |P_{\mathbf{j}}(\underline{\alpha})| \binom{j_1}{i_1} \cdots \binom{j_m}{i_m} \le (2C_1)^{md_1} \sum_{\mathbf{j}} 2^{j_1+\cdots+j_m} \le (6C_1)^{md_1}.$$

Hence, for

$$T(\mathbf{X}) := P_{\mathbf{i}}(\mathbf{X}) = \sum_{\mathbf{j}} T(\mathbf{j})(X_1 - \alpha)^{j_1} \cdots (X_m - \alpha)^{j_m}$$

we have

$$T(\mathbf{j}) = 0 \qquad \text{if} \qquad \frac{j_1}{d_1} + \cdots + \frac{j_m}{d_m} \le \frac{m}{2}(1 - 3\epsilon),$$

and

$$\sum_{\mathbf{j}} |T(\mathbf{j})| \le (6C_1)^{md_1}.$$

It follows that, on denoting by $*$ the $\mathbf{j}$ with $j_1/d_1 + \cdots + j_m/d_m > m(1 - 3\epsilon)/2$,

$$\begin{aligned}
|T(\frac{p_1}{q_1}, \ldots, \frac{p_m}{q_m})| &\le (6C_1)^{md_1} \max_{\mathbf{j}}^{*} \left|\frac{p_1}{q_1} - \alpha\right|^{j_1} \cdots \left|\frac{p_m}{q_m} - \alpha\right|^{j_m} \\
&\le (6C_1)^{md_1} \max_{\mathbf{j}}^{*} \left((q_1^{d_1})^{j_1/d_1} \cdots (q_m^{d_m})^{j_m/d_m}\right)^{-2-\delta} \qquad \text{(by (1.17))} \\
&\le (6C_1)^{md_1} \max_{\mathbf{j}}^{*} (q_1^{d_1})^{(j_1/d_1 + \cdots + j_m/d_m)(-2-\delta)} \qquad \text{(by (1.21))} \\
&\le q_1^{\epsilon m d_1} (q_1^{d_1})^{-m(1-3\epsilon)(1+\delta/2)} \qquad \text{(by (1.18))} \\
&\le (q_1^{d_1} \cdots q_m^{d_m})^{\{\epsilon - (1-3\epsilon)(1+\delta/2)\}/(1+\epsilon)} \qquad \text{(by (1.21))} \\
&< (q_1^{d_1} \cdots q_m^{d_m})^{-1} \qquad \text{(since } \delta = 12\epsilon < 1/2.)
\end{aligned}$$

On the other hand, $T(p_1/q_1, \ldots, p_m/q_m)$ is a rational number with denominator dividing $q_1^{d_1} \cdots q_m^{d_m}$. Thus

$$P_{\mathbf{i}}\left(\frac{p_1}{q_1}, \ldots, \frac{p_m}{q_m}\right) = T\left(\frac{p_1}{q_1}, \ldots, \frac{p_m}{q_m}\right) = 0 \qquad \text{if (1.22) holds.}$$

□

2 Variations and Generalisations

Roth's Theorem can be restated in the following symmetric form.

2.1 Theorem. *Let $L_1(X,Y) = \alpha X + \beta Y$ and $L_2(X,Y) = \gamma X + \delta Y$ be linearly independent linear forms with algebraic coefficients. Then, for every $\delta > 0$, the inequalities*

(2.2) $$0 < |L_1(x,y)L_2(x,y)| < \max(|x|,|y|)^{-\delta}$$

have only finitely many solutions in integers x, y.

Proof. Theorem 2.1 implies Roth's Theorem, since $|\alpha - x/y| < |y|^{-2-\delta}$ implies

$$|y(x-\alpha y)| < |y|^{-\delta} \ll (\max(|x|,|y|))^{-\delta}.$$

On the other hand, Roth's Theorem implies Theorem 2.1 by the following argument. Assume $|L_2(x,y)| \geq |L_1(x,y)|$. We have $|L_1(x,y)| \gg |y|\cdot|x/y+\beta/\alpha|$ and

$$|L_2(x,y)| \gg |\gamma L_1(x,y) - \alpha L_2(x,y)| = |\alpha\delta - \beta\gamma|\cdot|y| \gg |y|.$$

Hence

$$\left|-\frac{\beta}{\alpha} - \frac{x}{y}\right| \ll \frac{1}{|y|^2}|L_1(x,y)L_2(x,y)| < \frac{1}{|y|^{2+\delta}},$$

so there can be only finitely many solutions x/y. □

Ridout (1955) extended Roth's Theorem to p-adic numbers and LeVeque did so for approximations by elements from some fixed number field. In 1960 Lang formulated the following common generalisation.

2.3 Theorem. *Let K be an algebraic number field, S a finite set of places of K containing all infinite places. Let L be a finite extension of K. For each $v \in S$ choose a fixed extension of $|\ |_v$ (normalised valuation) to L. For $v \in S$, let $\alpha_v \in L$. Then the inequality*

$$\prod_{v\in S} \min(1, |\xi - \alpha_v|_v) < H_K(\xi)^{-2-\delta} \qquad \text{in } \xi \in K$$

has only finitely many solutions.

See S. Lang, [36], p. 160. Of course, there is also a symmetric form of Theorem 2.3. It says that for $(\alpha_v : \beta_v)$ in $\mathbb{P}^1(L)$ there are only finitely many $(\xi : \eta)$ in $\mathbb{P}^1(K)$ such that

$$\prod_{v\in S} \frac{|\alpha_v\xi + \beta_v\eta|_v}{\max(|\xi|_v, |\eta|_v)} < H_K(\xi:\eta)^{-2-\delta}.$$

This implies immediately that the S-unit equation

(2.4) $$x + y + z = 0 \qquad \text{in } x,\ y \text{ and } z \text{ in } \mathcal{O}_S^*$$

has only finitely many solutions. Indeed, for $v \in S$ choose $(\alpha_v : \beta_v) = (1:0)$, $(0:1)$ or $(-1:-1)$ according as $|x|_v$, $|y|_v$ or $|z|_v$ is the smallest of $|x|_v$, $|y|_v$ and $|z|_v$. Then

$$\prod_{v\in S} \frac{|\alpha_v x + \beta_v y|}{\max(|x|_v, |y|_v)} \ll \prod_{v\in S} \frac{|xyz|_v}{\max(|x|_v,|y|_v)^3} = \frac{1}{H_K(x,y)^3},$$

since $x, y, z \in O_S^*$, whence $|x|_v = |y|_v = |z|_v = 1$ for $v \notin S$. Thus there are only finitely many solutions $(x : y : z)$ of (2.4).

It is essential that in Theorem 2.3 the unknown ξ belongs to some fixed number field K. The 2 in the exponent is the best possible, as we see from the following result.

2.5 Theorem. *Let K be a real algebraic number field. Then there exists a constant c_K such that for every real α not in K there are infinitely many $\xi \in K$ having*

$$|\alpha - \xi| < c_k \max(1, |\alpha|^2) H_K(\xi)^{-2}.$$

See Schmidt, [65], p. 253. If only the degree of ξ is fixed, the exponent will depend on the degree.

2.6 Theorem. *Suppose α is a real algebraic number. Let $k \geq 1$, $\delta > 0$. Then there are only finitely many algebraic numbers ξ of degree $\leq k$ with*

$$|\alpha - \xi| < H(\xi)^{-k-1-\delta},$$

where H denotes the classical absolute height.

See Schmidt, [65], p. 278. The number $k + 1$ in the exponent is the best possible. Theorem 2.6 is a consequence of the Subspace Theorem and it therefore belongs to Chapter IV.

Acknowledgement. I am indebted to Dr. J.H. Evertse for making a draft for part of this text and for making some critical remarks.

Chapter IV

The Subspace Theorem of W.M. Schmidt

by Jan-Hendrik Evertse

1 Introduction

In Chapter III about Roth's theorem, the following equivalent formulation of this theorem was proved:

1.1 Theorem. *Let $l_1(X,Y) = \alpha X+\beta Y$, $l_2(X,Y) = \gamma X+\delta Y$ be linearly independent linear forms with (real or complex) algebraic coefficients. Then for every $\epsilon > 0$, the number of solutions of*

(1.2) $\quad |l_1(x,y)l_2(x,y)| < \max(|x|,|y|)^{-\epsilon} \qquad$ *in* $(x,y) \in \mathbb{Z}^2$ *with* $\gcd(x,y) = 1$

is finite.

A natural generalisation of (1.2) is the inequality

(1.3) $$|l_1(\mathbf{x})\cdots l_n(\mathbf{x})| < |\mathbf{x}|^{-\epsilon} \qquad \text{in } \mathbf{x} \in \mathbb{Z}^n,$$

where $|\mathbf{x}| = \max(|x_1|,\dots,|x_n|)$ for $\mathbf{x} = (x_1,\dots,x_n)$ in $\mathbb{Z}^n$ and where $l_1,\dots,l_n$ are linearly independent linear forms with algebraic coefficients. Inequalities of type (1.3) might have infinitely many solutions $(x_1,\dots,x_n)$ with $\gcd(x_1,\dots,x_n) = 1$. Consider for instance

(1.4) $$\left|(x_1 + \sqrt{2}x_2 + \sqrt{3}x_3)(x_1 + \sqrt{2}x_2 - \sqrt{3}x_3)(x_1 - \sqrt{2}x_2 - \sqrt{3}x_3)\right| < |\mathbf{x}|^{-\epsilon}$$

in $(x_1,x_2,x_3) \in \mathbb{Z}^3$, where $0 < \epsilon < 1$. It is easy to check that every triple $(x_1,x_2,x_3) \in \mathbb{Z}^3$ with $x_3 = 0$, $x_1 > 0$, $x_2 < 0$ and $x_1^2 - 2x_2^2 = 1$ satisfies (1.4). Hence (1.4) has infinitely many solutions with $\gcd(x_1,x_2,x_3) = 1$ in the subspace $x_3 = 0$. Similarly, (1.4) has infinitely many such solutions in the subspaces $x_1 = 0$ and $x_2 = 0$. It appears that the set of solutions of (1.4) with $x_1x_2x_3 \neq 0$ is contained in finitely many proper linear subspaces of $\mathbb{Q}^3$. Namely, one has

1.5 Theorem. *(Subspace Theorem, W.M. Schmidt, 1972 [69]). For every $\epsilon > 0$ there are a finite number of proper linear subspaces $T_1,\dots,T_h$ of $\mathbb{Q}^n$ such that the set of solutions of (1.3) is contained in $T_1 \cup \cdots \cup T_h$.*

Theorem 1.5 can be reformulated in a more geometrical way as follows. The *Subspace topology* (notion introduced by Schmidt) on $\mathbb{Q}^n$ is the topology of which the closed sets are the finite unions of linear subspaces of $\mathbb{Q}^n$. Then Theorem 1.5 states that for every $\epsilon > 0$, the set of solutions of (1.3) is not dense in $\mathbb{Q}^n$ w.r.t. the Subspace topology.

The Subspace Theorem has a generalisation over number fields, involving non-archimedean absolute values. The theorem below is equivalent to a result proved by Schlickewei [64]. The following notation is used:

> K is an algebraic number field; $\{\|\cdot\|_v : v \in M_K\}$ is a maximal set of pairwise inequivalent absolute values on K, normalised such that the Product Formula $\prod_v \|x\|_v = 1$ for $x \in K^*$ holds (cf. [14], p. 151); each $\|\cdot\|_v$ has been extended in some way to $\overline{\mathbb{Q}}$; $H_K(\mathbf{x}) = \prod_v \|\mathbf{x}\|_v$, where $\|\mathbf{x}\|_v = \max(\|x_1\|_v, \ldots, \|x_n\|_v)$ for $\mathbf{x} \in \mathbb{P}^{n-1}(K)$; S is a finite subset of M_K; $\{l_{1v}, \ldots, l_{nv}\}$ ($v \in S$) are linearly independent sets of linear forms in $\overline{\mathbb{Q}}[X_1, \ldots, X_n]$.

1.6 Theorem. *For every $\epsilon > 0$ there are a finite number $T_1, \ldots, T_h$ of proper hyperplanes of $\mathbb{P}^{n-1}(K)$ such that the set of solutions of*

$$\prod_{v \in S} \prod_{i=1}^{n} \frac{\|l_{iv}(\mathbf{x})\|_v}{\|\mathbf{x}\|_v} < H_K(\mathbf{x})^{-n-\epsilon} \qquad \text{in } \mathbf{x} \in \mathbb{P}^{n-1}(K) \tag{1.7}$$

is contained in $T_1 \cup \cdots \cup T_h$.

Note that (1.3) implies that $\prod_{i=1}^n (\|l_i(\mathbf{x})\|_\infty / \|\mathbf{x}\|_\infty) < H_{\mathbb{Q}}(\mathbf{x})^{-n-\epsilon}$ for $\mathbf{x} \in \mathbb{Z}^n$, where $\|\cdot\|_\infty$ is the usual absolute value. Hence Theorem 1.6 implies Theorem 1.5.

The following slight generalisation of Theorem 1.6, due to Vojta [80], is more convenient for certain applications. A set of linear forms $\{l_1, \ldots, l_m\}$ in $\overline{\mathbb{Q}}[X_1, \ldots, X_n]$ is said to be in *general position* if each subset of cardinality $\leq n$ of $\{l_1, \ldots, l_m\}$ is linearly independent. For $v \in S$, let $\{l_{1v}, \ldots, l_{m_v,v}\}$ be a set of linear forms in $\overline{\mathbb{Q}}[X_1, \ldots, X_n]$ in general position.

1.8 Theorem. *For every $\epsilon > 0$ there are a finite number $T_1, \ldots, T_h$ of proper hyperplanes of $\mathbb{P}^{n-1}(K)$ such that the set of solutions of*

$$\prod_{v \in S} \prod_{i=1}^{m_v} \frac{\|l_{iv}(\mathbf{x})\|_v}{\|\mathbf{x}\|_v} < H_K(\mathbf{x})^{-n-\epsilon} \qquad \text{in } \mathbf{x} \in \mathbb{P}^{n-1}(K) \tag{1.9}$$

is contained in $T_1 \cup \cdots \cup T_h$.

Proof. (Theorem 1.6 implies Theorem 1.8.) If $m_v < n$ then we may assume that $\{l_{1v}, \ldots, l_{m_v}, X_{m_v+1}, \ldots, X_n\}$ is linearly independent. Put $l'_{iv} = l_{iv}$ for $i \leq m_v$, $l'_{iv} = X_i$ for $i > m_v$. Thus, $\|l'_{iv}(\mathbf{x})\|_v \leq \|\mathbf{x}\|_v$ for $\mathbf{x} \in K^n$, $i > m_v$. Now suppose that $m_v \geq n$. Fix $\mathbf{x} \in K^n$. There is a permutation $(j_1, \ldots, j_{m_v})$ of $(1, \ldots, m_v)$ such that $\|l_{j_1}(\mathbf{x})\|_v \leq \|l_{j_2}(\mathbf{x})\|_v \leq \cdots \leq \|l_{j_{m_v}}(\mathbf{x})\|_v$. Put $l'_{iv} = l_{j_i,v}$ for $i = 1, \ldots, n$. For $i > n$, the set $\{l_{j_1}, \ldots, l_{j_{n-1}}, l_{j_i}\}$ is linearly independent. Hence for $i > n$, $k = 1, \ldots, n$ we have $X_k = \alpha_{ik1} l_{j_1} + \cdots + \alpha_{ik,n-1} l_{j_{n-1}} + \alpha_{ikn} l_{j_i}$ for suitable coefficients α_{ikh}. It follows that for $i > n$,

$$\|\mathbf{x}\|_v \ll \max(\|l_{j_1}(\mathbf{x})\|_v, \ldots, \|l_{j_i}(\mathbf{x})\|_v) = \|l_{j_i}(\mathbf{x})\|_v.$$

Hence every solution $\mathbf{x}$ of (1.9) satisfies one of a finite number of inequalities

$$\prod_{v\in S}\prod_{i=1}^{n}\frac{\|l'_{iv}(\mathbf{x})\|_v}{\|\mathbf{x}\|_v} \ll H_K(\mathbf{x})^{-n-\epsilon}.$$

By applying Theorem 1.6 to each of these inequalities we obtain Theorem 1.8. □

2 Applications

In 1842, Dirichlet proved the following result:

> *Let $\{\alpha_1,\dots,\alpha_n\}$ be a $\mathbb{Q}$-linearly independent set of real numbers. Then for some $C>0$, the inequality*
>
> $$|\alpha_1x_1+\dots+\alpha_nx_n| < C|\mathbf{x}|^{-n+1} \qquad \text{in } \mathbf{x}\in\mathbb{Z}^n$$
>
> *has infinitely many solutions.*

In 1970, Schmidt ([68], see also [65], p. 152) proved that the exponent $-n+1$ cannot be replaced by a smaller number if $\alpha_1,\dots,\alpha_n$ are algebraic.

2.1 Theorem. *Let $\alpha_1,\dots,\alpha_n$ be algebraic numbers. Then for every $\epsilon>0$, the inequality*

(2.2) $$0<|\alpha_1x_1+\dots+\alpha_nx_n|<|\mathbf{x}|^{-n+1-\epsilon} \qquad \text{in } \mathbf{x}\in\mathbb{Z}^n$$

has only finitely many solutions.

Proof. Induction on n. For $n=1$, the assertion is trivial. Let $n>1$. We may assume that $\alpha_1\neq 0$. Then the set of linear forms $\{\alpha_1X_1+\dots+\alpha_nX_n, X_2,\dots,X_n\}$ is linearly independent. For every solution $\mathbf{x}$ of (2.2) we have

(2.3) $$|(\alpha_1x_1+\dots+\alpha_nx_n)x_2\cdots x_n|<|\mathbf{x}|^{-\epsilon}.$$

By Theorem 1.5, the set of solutions of (2.3), and hence (2.2), is contained in finitely many proper linear subspaces of $\mathbb{Q}^n$. Let T be one of these subspaces. Without loss of generality we may assume that $\alpha_1X_1+\dots+\alpha_nX_n=\beta_1X_1+\dots+\beta_{n-1}X_{n-1}$ identically on T, where $\beta_1,\dots,\beta_{n-1}$ are algebraic numbers. Hence for every solution $\mathbf{x}$ of (2.3) with $\mathbf{x}\in T$ we have

(2.4) $$0<|\beta_1x_1+\dots+\beta_{n-1}x_{n-1}|<|\mathbf{x}|^{-n+1-\epsilon}<\max(|x_1|,\dots,|x_{n-1}|)^{-n+2-\epsilon}.$$

By the induction hypothesis, (2.4) has only finitely many solutions in $(x_1,\dots,x_{n-1})\in\mathbb{Z}^{n-1}$. Hence (2.2) has only finitely many solutions with $\mathbf{x}\in T$. Since there are only finitely many possibilities for T this completes the induction step. □

From Theorem 2.1, one can derive the following generalisation of Roth's theorem. Here, the *height* $H(\xi)$ of ξ is the maximum of the absolute values of the coefficients of the minimal polynomial of ξ.

2.5 Theorem. *For every algebraic number $\alpha\in\mathbb{C}$, integer $d\geq 1$, and $\epsilon>0$, there are only finitely many algebraic numbers $\xi\in\mathbb{C}$ of degree d such that*

(2.6) $$|\alpha-\xi|<H(\xi)^{-d-1-\epsilon}.$$

Proof. Let ξ be an algebraic number of degree d satisfying (2.6). We may assume that ξ is not a conjugate of α. Denote by $f(X) = x_{d+1}X^d + x_d X^{d-1} + \cdots + x_1 \in \mathbb{Z}[X]$ the minimal polynomial of ξ. Then $H(\xi) = |\mathbf{x}|$ and $f(\alpha) \neq 0$. From, e.g., the mean value theorem it follows that $|f(\alpha)|/|\alpha - \xi| \ll_\alpha H(\xi)$. Hence

$$(2.7) \qquad \begin{aligned} 0 < |x_1 + x_2\alpha + \cdots + x_{d+1}\alpha^d| = |f(\alpha)| \ll_\alpha |\alpha - \xi| H(\xi) < \\ < H(\xi)^{-d-\epsilon} = |\mathbf{x}|^{-d-\epsilon}. \end{aligned}$$

By Theorem 2.1, there are only finitely many $\mathbf{x} \in \mathbb{Z}^{d+1}$ satisfying (2.7). This implies that there are only finitely many ξ of degree d with $|\alpha - \xi| < H(\xi)^{-d-1-\epsilon}$. □

We consider the equation

$$(2.8) \qquad a_1 x_1 + \cdots + a_n x_n = 1 \qquad \text{in } x_1, \ldots, x_n \in G,$$

where G is a finitely generated multiplicative subgroup of $\overline{\mathbb{Q}}^*$ and $a_1, \ldots, a_n \in \overline{\mathbb{Q}}^*$. A solution $(x_1, \ldots, x_n)$ of (2.8) is called *non-degenerate* if $\sum_{i \in I} a_i x_i \neq 0$ for each non-empty subset I of $\{1, \ldots, n\}$.

2.9 Theorem. *([18], [59]). Equation (2.8) has only finitely many non-degenerate solutions.*

2.10 Lemma. *There is a finite set U such that for every solution (either or not non-degenerate) of (2.8) there is an $i \in \{1, \ldots, n\}$ with $x_i \in U$.*

Proof. Induction on n. For $n = 1$, the assertion is trivial. Let $n > 1$. There are an algebraic number field K and a finite set of places S of K, containing all archimedean places, such that G is contained in the group of S-units $\{x \in K : \|x\|_v = 1 \text{ for } v \notin S\}$. Define the linear form $X_0 := a_1 X_1 + \cdots + a_n X_n$. Then $\{X_0, X_1, \ldots, X_n\}$ is in general position. Let $\mathbf{x} = (x_1, \ldots, x_n)$ be a solution of (2.8) and put $x_0 := 1$. Then, from the fact that $x_0, \ldots, x_n$ are S-units and from the Product Formula, it follows that $\prod_{v \in S} \|x_0 x_1 \cdots x_n\|_v = 1$. Further, $H_K(\mathbf{x}) = \prod_{v \in S} \|\mathbf{x}\|_v$. Hence

$$\prod_{v \in S} \frac{\|x_0 \cdots x_n\|_v}{\|\mathbf{x}\|_v^{n+1}} = H_K(\mathbf{x})^{-n-1}.$$

Now Theorem 1.8 implies that $\mathbf{x} \in T_1 \cup \cdots \cup T_h$ where $T_1, \ldots, T_h$ are proper hyperplanes of $\mathbb{P}^{n-1}(K)$ independent of $\mathbf{x}$. Now let $\mathbf{x}$ be a solution of (2.8) lying in a proper projective subspace T of $\mathbb{P}^{n-1}(K)$ defined by an equation $b_1 X_1 + \cdots + b_n X_n = 0$. By combining this with (2.8) we obtain $\sum_{i \in J} c_i x_i = 1$, where c_i $(i \in J)$ are elements of K^* depending only on $a_1, \ldots, a_n$, T, and J is a proper subset of $\{1, \ldots, n\}$. Now the induction hypothesis implies that $x_i \in U_T$ for some $i \in J$, where U_T is a finite set depending on $a_1, \ldots, a_n$ and T. Then clearly, for every solution $(x_1, \ldots, x_n)$ of (2.8) there is an i with $x_i \in U := U_{T_1} \cup \cdots \cup U_{T_h}$. □

Proof. (of Theorem 2.9.) Induction on n. For $n = 1$, the assertion is again trivial. Let $n > 1$. In view of Lemma 2.10, it suffices to show that (2.8) has only finitely many non-degenerate solutions with $x_n = c$, where $c \in U$ is fixed. A solution with $x_n = a_n^{-1}$ is degenerate so we may assume that $a_n c \neq 1$. Then clearly, $(x_1, \ldots, x_{n-1})$ is a non-degenerate solution of $(1 - a_n c)^{-1} a_1 x_1 + \cdots + (1 - a_n c)^{-1} a_{n-1} x_{n-1} = 1$. By the induction hypothesis, there are only finitely many possibilities for $x_1, \ldots, x_{n-1}$.
□

We now prove a generalisation of Theorem 2.9 which was part of a conjecture of Lang (cf. [36], p. 220), and proved by Laurent [41]. The product of $\mathbf{x} = (x_1, \ldots, x_n)$ and $\mathbf{y} = (y_1, \ldots, y_n)$ in $(\overline{\mathbb{Q}}^*)^n$ is defined by $\mathbf{x} \cdot \mathbf{y} = (x_1y_1, \ldots, x_ny_n)$. We may view $(\overline{\mathbb{Q}}^*)^n$ with this coordinatewise multiplication as an algebraic group (a so-called *linear torus*). For every n-tuple $\mathbf{i} = (i_1, \ldots, i_n)$ in $(\mathbb{Z}_{\geq 0})^n$ we write $\mathbf{X}^{\mathbf{i}} = X_1^{i_1} \cdots X_n^{i_n}$. An *algebraic subgroup* of $(\overline{\mathbb{Q}}^*)^n$ is a set

$$H = \{\mathbf{x} \in (\overline{\mathbb{Q}}^*)^n : \; f_1(\mathbf{x}) = 0, \ldots, f_t(\mathbf{x}) = 0\} \qquad (f_1, \ldots, f_t \in \overline{\mathbb{Q}}[X_1, \ldots, X_n])$$

which is closed under coordinatewise multiplication and inversion. It is not difficult to prove that every algebraic subgroup is of the form

$$\{\mathbf{x} \in (\overline{\mathbb{Q}}^*)^n : \mathbf{x}^{\mathbf{i}} = \mathbf{x}^{\mathbf{j}} \qquad \text{for } (\mathbf{i}, \mathbf{j}) \in I\},$$

where I is a finite set of pairs from $(\mathbb{Z}_{\geq 0})^n$. We shall not need this fact.

2.11 Theorem. *([41]) Let V be an algebraic subvariety of $(\overline{\mathbb{Q}}^*)^n$ and G a finitely generated subgroup of $(\overline{\mathbb{Q}}^*)^n$. Then $V \cap G$ is a finite union of "families" $\mathbf{u} \cdot (H \cap G)$, where H is an algebraic subgroup of $(\overline{\mathbb{Q}}^*)^n$ and $\mathbf{u} \in G$ is such that $\mathbf{u} \cdot H \subset V$.*

2.12 Remark. Faltings [23] proved the other part of Lang's conjecture on p. 220 of [36], which is the analogue of Theorem 2.11 for abelian varieties, see Chapter XIII. In this analogue, one has to take instead of $(\overline{\mathbb{Q}}^*)^n$ an abelian variety A over an algebraic number field K and instead of G the group of K-rational points $A(K)$ of K. □

2.13 Remark. Laurent [41] proved in fact a more general result than Theorem 2.11, in which V is a subvariety of $(\mathbb{C}^*)^n$ and G a subgroup of $(\mathbb{C}^*)^n$ of finite rank, i.e., G has a finitely generated subgroup G_0 such that G/G_0 is a torsion group. □

Proof. (of Theorem 2.11.) Theorem 2.11 can be derived from Theorem 2.9 by using combinatorics. We have

$$V = \{\mathbf{x} \in (\overline{\mathbb{Q}}^*)^n : f_1(\mathbf{x}) = 0, \ldots, f_t(\mathbf{x}) = 0\},$$

where

$$f_k(\mathbf{X}) = \sum_{\mathbf{i} \in C_k} a(\mathbf{i}, k)\mathbf{X}^{\mathbf{i}},$$

with C_k being a finite subset of $(\mathbb{Z}_{\geq 0})^n$ and $a(\mathbf{i}, k)$ in $\overline{\mathbb{Q}}^*$, for $k = 1, \ldots, t, \mathbf{i}$ in C_k. Let $\mathcal{P}$ denote the collection of subsets of cardinality 2 of $C_1 \cup \cdots \cup C_t$. To every $\mathbf{x} \in V$ we associate a subcollection $\mathcal{E}_{\mathbf{x}}$ of $\mathcal{P}$, which consists of the sets $\{\mathbf{p}, \mathbf{q}\}$ with the following property:

there are $k \in \{1, \ldots, t\}$ and a subset C of C_k such that

$$\mathbf{p}, \mathbf{q} \in C,$$
$$\sum_{\mathbf{i} \in C} a(\mathbf{i}, k)\mathbf{x}^{\mathbf{i}} = 0,$$
$$\sum_{\mathbf{i} \in C'} a(\mathbf{i}, k)\mathbf{x}^{\mathbf{i}} \neq 0 \qquad \text{for each proper, non-empty subset } C' \text{ of } C.$$

For each subcollection $\mathcal{E}$ of $\mathcal{P}$, put

$$V(\mathcal{E}) = \{\mathbf{x} \in V : \mathcal{E}_{\mathbf{x}} = \mathcal{E}\}$$

and define the algebraic subgroup of $(\overline{\mathbb{Q}}^*)^n$:

$$H(\mathcal{E}) = \{\mathbf{x} \in (\overline{\mathbb{Q}}^*)^n : \mathbf{x}^{\mathbf{i}} = \mathbf{x}^{\mathbf{j}} \quad \text{for each } \{\mathbf{i},\mathbf{j}\} \text{ in } \mathcal{E}\}.$$

2.14 Lemma. *Let $\mathcal{E} \subseteq \mathcal{P}$ and $\mathbf{u} \in V(\mathcal{E})$. Then $\mathbf{u} \cdot H(\mathcal{E}) \subseteq V$.*

Proof. There are pairwise disjoint subsets $E_1, \ldots, E_s$ of C_1 such that $C_1 = E_1 \cup \cdots \cup E_s$ and such that for $l = 1, \ldots, s$,

$$\sum_{\mathbf{i} \in E_l} a(\mathbf{i},1)\mathbf{u}^{\mathbf{i}} = 0, \qquad \sum_{\mathbf{i} \in E'} a(\mathbf{i},1)\mathbf{u}^{\mathbf{i}} \neq 0 \qquad \text{for each nonempty } E' \subsetneqq E_l.$$

Since $\mathbf{u} \in V(\mathcal{E})$ we have $\mathcal{E}_{\mathbf{u}} = \mathcal{E}$. Hence if $\mathbf{i}$ and $\mathbf{j}$ belong to the same set E_l, then $\{\mathbf{i},\mathbf{j}\} \in \mathcal{E}$ which implies that $\mathbf{x}^{\mathbf{i}} = \mathbf{x}^{\mathbf{j}}$ for every $\mathbf{x} \in H(\mathcal{E})$. Therefore, for every $\mathbf{x} \in H(\mathcal{E})$ we have

$$\sum_{\mathbf{i} \in E_l} a(\mathbf{i},1)(\mathbf{u}\cdot\mathbf{x})^{\mathbf{i}} = \mathbf{x}^{\mathbf{i}_l} \sum_{\mathbf{i} \in E_l} a(\mathbf{i},1)\mathbf{u}^{\mathbf{i}} = 0 \qquad \text{for } l = 1, \ldots, s,$$

where $\mathbf{i}_l$ is a fixed tuple from E_l. By taking the sum over all l we get $f_1(\mathbf{u}\cdot\mathbf{x}) = 0$ for all $\mathbf{x} \in H(\mathcal{E})$. Similarly, $f_2(\mathbf{u}\cdot\mathbf{x}) = \cdots = f_t(\mathbf{u}\cdot\mathbf{x}) = 0$ for all $\mathbf{x} \in H(\mathcal{E})$. Hence $\mathbf{u} \cdot H(\mathcal{E}) \subseteq V$. □

2.15 Lemma. *For each subcollection $\mathcal{E}$ of $\mathcal{P}$ there is a finite subset $W(\mathcal{E})$ of $V(\mathcal{E}) \cap G$ such that*

$$V(\mathcal{E}) \cap G \subseteq \bigcup_{\mathbf{u} \in W(\mathcal{E})} \mathbf{u} \cdot (H(\mathcal{E}) \cap G).$$

Proof. Theorem 2.9 can be reformulated as follows: Let $a_0, \ldots, a_n \in \overline{\mathbb{Q}}^*$ and let G_0 be a finitely generated subgroup of $\overline{\mathbb{Q}}^*$. Then there is a finite set U, such that $x_i/x_j \in U$ for every non-degenerate solution $(x_0, \ldots, x_n)$ in G_0^{n+1} of $a_0x_0 + \cdots + a_nx_n = 0$ and for all $i, j \in \{0, \ldots, n\}$. Now let $\mathbf{x} \in V(\mathcal{E}) \cap G$ and take $\{\mathbf{p},\mathbf{q}\} \in \mathcal{E}$. Choose a set C with $\mathbf{p},\mathbf{q} \in C$ and $C \subseteq C_k$ for some $k \in \{1, \ldots, t\}$ as in the definition of $\mathcal{E}_{\mathbf{x}}$. By applying the above reformulation of Theorem 2.9 to $\sum_{\mathbf{i} \in C} a(\mathbf{i},k)\mathbf{x}^{\mathbf{i}} = 0$, we infer that $\mathbf{x}^{\mathbf{p}}/\mathbf{x}^{\mathbf{q}} \in U$, where U is some finite set, depending only on $f_1, \ldots, f_t, G$.

We can choose the same set U for each $\{\mathbf{p},\mathbf{q}\} \in \mathcal{E}$. Thus,

$$\mathbf{x}^{\mathbf{p}}/\mathbf{x}^{\mathbf{q}} \in U \qquad \text{for each } \{\mathbf{p},\mathbf{q}\} \in \mathcal{E}.$$

This implies that there is a finite subset $W(\mathcal{E})$ of $V(\mathcal{E}) \cap G$ such that for every $\mathbf{x} \in V(\mathcal{E}) \cap G$ there is an $\mathbf{u} \in W(\mathcal{E})$ with $\mathbf{x}^{\mathbf{p}}/\mathbf{x}^{\mathbf{q}} = \mathbf{u}^{\mathbf{p}}/\mathbf{u}^{\mathbf{q}}$ for each $\{\mathbf{p},\mathbf{q}\} \in \mathcal{E}$, in other words, $(\mathbf{x}/\mathbf{u})^{\mathbf{p}} = (\mathbf{x}/\mathbf{u})^{\mathbf{q}}$ for each $\{\mathbf{p},\mathbf{q}\} \in \mathcal{E}$ or $\mathbf{x} \in \mathbf{u}\cdot(H(\mathcal{E}) \cap G)$. This implies Lemma 2.15. □

Lemma 2.15 implies that

$$V \cap G \subseteq \bigcup_{\mathcal{E}} \bigcup_{\mathbf{u} \in W(\mathcal{E})} \mathbf{u} \cdot (H(\mathcal{E}) \cap G),$$

where the union is taken over all subcollections $\mathcal{E}$ of $\mathcal{P}$. By Lemma 2.14, these two sets are equal. This completes the proof of Theorem 2.11. □

3 About the Proof of the Subspace theorem

We give an overview of the proof of Theorem 1.5. For certain details, we refer to [65], Chaps. V, VI. We remark that it suffices to prove Theorem 1.5 for the case that the linear forms in the inequality

$$|l_1(\mathbf{x})\cdots l_n(\mathbf{x})| < |\mathbf{x}|^{-\epsilon} \quad \text{in } \mathbf{x} \in \mathbb{Z}^n \tag{3.1}$$

have real algebraic integer coefficients. Namely, if for instance some of the coefficients of l_n are complex, then write $l_n = l_{n1} + \sqrt{-1}l_{n2}$, where l_{n1}, l_{n2} are linear forms with real algebraic coefficients. Choose l'_n from $\{l_{n1}, l_{n2}\}$ such that $\{l_1, \ldots, l_{n-1}, l'_n\}$ is linearly independent. Then $|l'_n(\mathbf{x})| \leq |l_n(\mathbf{x})|$ for $\mathbf{x} \in \mathbb{Z}^n$. Hence (3.1) remains valid if we replace l_n by l'_n. Similarly, we can replace $l_1, \ldots, l_{n-1}$ by linear forms with real algebraic coefficients. Thus, we can reduce (3.1) to a similar inequality with linearly independent linear forms $l'_1, \ldots, l'_n$ with real algebraic coefficients. Choose a positive rational integer a such that the linear forms $l''_i := al'_i$ $(i = 1, \ldots, n)$ have real algebraic integer coefficients. Then for $|\mathbf{x}|$ sufficiently large, we get $|l''_1(\mathbf{x})\cdots l''_n(\mathbf{x})| < |\mathbf{x}|^{-\epsilon/2}$.

3.2 Reduction to a Statement about Parallelepipeds

In what follows, we assume that $l_1, \ldots, l_n$ are linearly independent linear forms in n variables with real algebraic integer coefficients. For each tuple $\mathbf{A} = (A_1, \ldots, A_n)$ of positive reals, we consider the parallelepiped

$$\Pi(\mathbf{A}) = \{\mathbf{x} \in \mathbb{R}^n \colon |l_i(\mathbf{x})| \leq A_i \text{ for } i = 1, \ldots, n\}.$$

Put

$$Q(\mathbf{A}) := \max(A_1, \ldots, A_n, A_1^{-1}, \ldots, A_n^{-1}).$$

Theorem 1.5 follows from the following result about the parallelepipeds $\Pi(\mathbf{A})$:

3.2.1 Theorem. *For every $\epsilon > 0$, there are finitely many proper linear subspaces $T_1, \ldots, T_h$ of $\mathbb{Q}^n$ such that for all tuples $\mathbf{A} = (A_1, \ldots, A_n)$ of positive reals with*

$$A_1 A_2 \cdots A_n < Q(\mathbf{A})^{-\epsilon}, \tag{3.2.2}$$

and $Q(\mathbf{A})$ sufficiently large, the set $\Pi(\mathbf{A}) \cap \mathbb{Z}^n$ is contained in one of the spaces $T_1, \ldots, T_h$.

In the sequel, constants implied by the Vinogradov symbols $\ll$ and $\gg$ depend only on $l_1, \ldots, l_n, \epsilon$. By $a \ll\gg b$ we mean $a \ll b$ and $a \gg b$.

Proof. (Theorem 3.2.1 implies Theorem 1.5) Let $\mathbf{x}$ be a solution of (3.1) with $l_1(\mathbf{x})\cdots l_n(\mathbf{x}) \neq 0$. Put

$$A_i = |l_i(\mathbf{x})| \qquad \text{for } i = 1, \ldots, n.$$

Then

$$\mathbf{x} \in \Pi(\mathbf{A}) \cap \mathbb{Z}^n, \qquad A_1 \cdots A_n < |\mathbf{x}|^{-\epsilon}.$$

We estimate $Q(\mathbf{A})$ from above. Suppose that the number field K generated by the coefficients of $l_1, \ldots, l_n$ has degree D. On the one hand, we have $|l_i(\mathbf{x})| \ll |\mathbf{x}|$. On the other hand,

$$|l_i(\mathbf{x})| = \frac{|N_{K/\mathbb{Q}}(l_i(\mathbf{x}))|}{|l_i^{(2)}(\mathbf{x})| \cdots |l_i^{(D)}(\mathbf{x})|} \gg |\mathbf{x}|^{-D},$$

where $l_i^{(2)}(\mathbf{x}), \ldots, l_i^{(D)}(\mathbf{x})$ are the conjugates of $l_i(\mathbf{x})$ over $\mathbb{Q}$. Hence for $|\mathbf{x}|$ sufficiently large we have $Q(\mathbf{A}) \leq |\mathbf{x}|^{2D}$. Therefore,

$$A_1 \cdots A_n < Q(\mathbf{A})^{-\epsilon/2D}.$$

Moreover,

(3.2.3) $$|\mathbf{x}| \ll \max(|l_1(\mathbf{x})|, \ldots, |l_n(\mathbf{x}|) \leq Q(\mathbf{A}),$$

hence $Q(\mathbf{A})$ is large if $|\mathbf{x}|$ is large. Now Theorem 3.2.1 implies that $\mathbf{x} \in \Pi(\mathbf{A}) \cap \mathbb{Z}^n \subset T_1 \cup \cdots \cup T_h$ for certain proper linear subspaces $T_1, \ldots, T_h$ of $\mathbb{Q}^n$ independent of $\mathbf{x}$. □

3.2.4 Remark. If (3.2.2) holds and $Q(\mathbf{A})$ is sufficiently large, then $\operatorname{rank}(\Pi(\mathbf{A}) \cap \mathbb{Z}^n) \leq n-1$, where $\operatorname{rank}(\Pi(\mathbf{A}) \cap \mathbb{Z}^n)$ is the maximal number of linearly independent vectors in $\Pi(\mathbf{A}) \cap \mathbb{Z}^n$. Namely, take $\mathbf{x}_1, \ldots, \mathbf{x}_n$ in $\Pi(\mathbf{A}) \cap \mathbb{Z}^n$. Then

$$\begin{aligned} |\det(\mathbf{x}_1, \ldots, \mathbf{x}_n)| &= \frac{|\det(l_i(\mathbf{x}_j))|}{|\det(l_1, \ldots, l_n)|} \ll \max_\sigma |l_1(\mathbf{x}_{\sigma(1)}) \cdots l_n(\mathbf{x}_{\sigma(n)})| \\ &\leq A_1 \cdots A_n < Q(\mathbf{A})^{-\epsilon}, \end{aligned}$$

where the maximum is taken over all permutations σ of $\{1, \ldots, n\}$. Since $Q(\mathbf{A})$ is sufficiently large, this implies that $|\det(\mathbf{x}_1, \ldots, \mathbf{x}_n)| < 1$. But this number is a rational integer, hence must be 0. Therefore, $\{\mathbf{x}_1, \ldots, \mathbf{x}_n\}$ is linearly dependent. □

Analogous to Roth's theorem, we could try to prove Theorem 3.2.1 as follows. For m sufficiently large, we can construct a polynomial $P(\mathbf{X}_1, \ldots, \mathbf{X}_m)$ in m blocks of n variables, with rational integral coefficients, which is divisible by high powers of $l_i(\mathbf{X}_h)$, for $i = 1, \ldots, n$ and $h = 1, \ldots, m$. Suppose that Theorem 3.2.1 is not true. Then for every m we can choose $\mathbf{A}_1, \ldots, \mathbf{A}_m$ such that the sets $\Pi(\mathbf{A}_h)) \cap \mathbb{Z}^n$ $(h = 1, \ldots, m)$ are in some kind of general position. Prove that $|P_{\mathbf{i}}(\mathbf{x}_1, \ldots, \mathbf{x}_m)| < 1$ for all $\mathbf{x}_h \in \Pi(\mathbf{A}_h) \cap \mathbb{Z}^n$ $(h = 1, \ldots, m)$ and all partial derivatives $P_{\mathbf{i}}$ of P of small order. Then for these $\mathbf{x}_1, \ldots, \mathbf{x}_m$ and partial derivatives we have $P_{\mathbf{i}}(\mathbf{x}_1, \ldots, \mathbf{x}_m) = 0$. Suppose we were able to prove that on the other hand, $P_{\mathbf{i}}(\mathbf{x}_1, \ldots, \mathbf{x}_m) \neq 0$ for some $\mathbf{x}_h \in \Pi(\mathbf{A}_h) \cap \mathbb{Z}^n$ $(h = 1, \ldots, m)$ and some small order partial derivative. Thus, we would arrive at a contradiction.

Unfortunately, as yet such a non-vanishing result has been proved only for the special case that $\operatorname{rank}(\Pi(\mathbf{A}_h) \cap \mathbb{Z}^n) = n-1$ for $h = 1, \ldots, m$. Therefore, we proceed as follows. In Step 1 of the proof of Theorem 3.2.1 we show that it suffices to prove Theorem 3.2.1 for parallelepipeds $\Pi(\mathbf{A})$ with $\operatorname{rank}(\Pi(\mathbf{A}) \cap \mathbb{Z}^n) = n-1$, by constructing from each $\Pi(\mathbf{A})$ with $\operatorname{rank}(\Pi(\mathbf{A}) \cap \mathbb{Z}^n) < n-1$ a parallelepiped $\Pi'(\mathbf{B})$ in $\mathbb{Z}^N$ with $\operatorname{rank}(\Pi'(\mathbf{B}) \cap \mathbb{Z}^N) = N-1$, where in general $N > n$. For this, we need geometry of numbers. Then, in Step 2 we prove Theorem 3.2.1 for parallelepipeds $\Pi(\mathbf{A})$ with $\operatorname{rank}(\Pi(\mathbf{A}) \cap \mathbb{Z}^n) = n-1$ in the way described above.

3.3 Step 1 of the Proof of Theorem 3.2.1

If $\mathrm{rank}(\Pi(\mathbf{A}) \cap \mathbb{Z}^n) = r < n-1$, then $\Pi'(\mathbf{B})$ will be contained in the exterior product $\wedge^{n-k}(\mathbb{R}^n) \cong \mathbb{R}^{\binom{n}{k}}$, for some k with $r \le k \le n-1$. We need to recall some facts about exterior products.

Put $N := \binom{n}{k}$. Let $\sigma_1, \dots, \sigma_N$ denote the subsets of cardinality $n-k$ of $\{1, \dots, n\}$, ordered lexicographically: thus, $\sigma_1 = \{1, \dots, n-k\}$, $\sigma_2 = \{1, \dots, n-k-1, n-k+1\}, \dots, \sigma_{N-1} = \{k, k+2, \dots, n\}$, $\sigma_N = \{k+1, k+2, \dots, n\}$. Let $\{\mathbf{e}_1 = (1, 0, \dots, 0), \mathbf{e}_2, \dots, \mathbf{e}_n\}$ be the standard basis of $\mathbb{R}^n$ and $\{\mathbf{E}_1, \dots, \mathbf{E}_N\}$ the standard basis of $\mathbb{R}^N$. The multilinear mapping from $\mathbb{R}^n \times \dots \times \mathbb{R}^n$ ($n-k$ times) to $\mathbb{R}^N$ sending $(\mathbf{x}_1, \dots, \mathbf{x}_{n-k})$ to $\mathbf{x}_1 \wedge \dots \wedge \mathbf{x}_{n-k}$ is defined by

- $\mathbf{e}_{i_1} \wedge \dots \wedge \mathbf{e}_{i_{n-k}} = \mathbf{0}$ if $i_1, \dots, i_{n-k}$ are not all distinct;
- $\mathbf{e}_{i_1} \wedge \dots \wedge \mathbf{e}_{i_{n-k}} = \pm \mathbf{E}_i$ where $\{i_1, \dots, i_{n-k}\} = \sigma_i$, the sign is + if the permutation needed to rearrange $i_1, \dots, i_{n-k}$ into increasing order is even, and the sign is − if this permutation is odd.

It is easy to verify that if $\{\mathbf{a}_1, \dots, \mathbf{a}_n\}$ and $\{\mathbf{b}_1, \dots, \mathbf{b}_n\}$ are two bases of $\mathbb{R}^n$ related by $\mathbf{b}_i = \sum_{j=1}^n \xi_{ij} \mathbf{a}_j$ for $i = 1, \dots, n$, then $\{\mathbf{A}_1, \dots, \mathbf{A}_N\}$, and $\{\mathbf{B}_1, \dots, \mathbf{B}_N\}$ with $\mathbf{A}_i = \mathbf{a}_{i_1} \wedge \dots \wedge \mathbf{a}_{i_{n-k}}$ and $\mathbf{B}_i = \mathbf{b}_{i_1} \wedge \dots \wedge \mathbf{b}_{i_{n-k}}$ where $\sigma_i = \{i_1 < \dots < i_{n-k}\}$, are two bases of $\mathbb{R}^N$ related by:

$$\mathbf{B}_i = \sum_{j=1}^{N} \Xi_{ij} \mathbf{A}_j \qquad \text{for } i = 1, \dots, N, \text{ with } \Xi_{ij} = \det(\xi_{pq})_{p \in \sigma_i, q \in \sigma_j}.$$

(We write $\sigma_i = \{i_1 < \dots < i_{n-k}\}$ if $\sigma_i = \{i_1, \dots, i_{n-k}\}$ and $i_1 < \dots < i_{n-k}$.) We need the following fact:

3.3.1 Lemma. *Let $k \in \{1, \dots, n\}$. If $\{\mathbf{a}_1, \dots, \mathbf{a}_n\}$ and $\{\mathbf{b}_1, \dots, \mathbf{b}_n\}$ are two bases of $\mathbb{R}^n$, then $\{\mathbf{a}_1, \dots, \mathbf{a}_k\}$ and $\{\mathbf{b}_1, \dots, \mathbf{b}_k\}$ generate the same vector space if and only if $\{\mathbf{A}_1, \dots, \mathbf{A}_{N-1}\}$ and $\{\mathbf{B}_1, \dots, \mathbf{B}_{N-1}\}$ generate the same vector space.*

Proof. Use that $\{\mathbf{a}_1, \dots, \mathbf{a}_k\}$ and $\{\mathbf{b}_1, \dots, \mathbf{b}_k\}$ generate the same space $\Longleftrightarrow \Xi_{iN} = 0$ for $i = 1, \dots, N-1 \Longleftrightarrow \{\mathbf{A}_1, \dots, \mathbf{A}_{N-1}\}$ and $\{\mathbf{B}_1, \dots, \mathbf{B}_{N-1}\}$ generate the same space. □

Lemma 3.3.1 implies that there is an *injective* map f_k from the collection of k-dimensional subspaces of $\mathbb{R}^n$ to the collection of $(N-1)$-dimensional subspaces of $\mathbb{R}^N$ such that if V has basis $\{\mathbf{a}_1, \dots, \mathbf{a}_k\}$ and $\mathbf{a}_{k+1}, \dots, \mathbf{a}_n$ are any vectors such that $\{\mathbf{a}_1, \dots, \mathbf{a}_n\}$ is a basis of $\mathbb{R}^n$, then $\{\mathbf{A}_1, \dots, \mathbf{A}_{N-1}\}$ is a basis of $f_k(V)$.

Now let $l_1, \dots, l_n$ be linearly independent linear forms in n variables with real coefficients (at this point, the coefficients need not be algebraic). Write $l_i(\mathbf{X}) = (\mathbf{c}_i, \mathbf{X})$ (usual scalar product in $\mathbb{R}^n$), put $\mathbf{C}_i = \mathbf{c}_{i_1} \wedge \dots \wedge \mathbf{c}_{i_{n-k}}$ for $i = 1, \dots, N$ where $i_1 < \dots < i_{n-k}$ and $\{i_1, \dots, i_{n-k}\} = \sigma_i$, and define the linear forms on $\mathbb{R}^N$:

$$L_i(\mathbf{X}) = (\mathbf{C}_i, \mathbf{X}) \qquad (i = 1, \dots, N).$$

Note that $L_1, \dots, L_N$ are linearly independent. For every $N = \binom{n}{k}$-tuple $\mathbf{B} = (B_1, \dots, B_N)$ of positive reals, define

$$\begin{aligned} \Pi_k(\mathbf{B}) &= \{\mathbf{x} \in \mathbb{R}^N : |L_i(\mathbf{x})| \le B_i \text{ for } i = 1, \dots, N\}, \\ Q(\mathbf{B}) &= \max\{B_1, \dots, B_N, B_1^{-1}, \dots, B_N^{-1}\}. \end{aligned}$$

3.3.2 Proposition. *For every $\epsilon > 0$ and for every tuple $\mathbf{A} = (A_1, \ldots, A_n)$ of positive reals with $Q(\mathbf{A})$ sufficiently large and*

$$A_1 \cdots A_n < Q(\mathbf{A})^{-\epsilon}, \qquad 1 \le \operatorname{rank}(\Pi(\mathbf{A}) \cap \mathbb{Z}^n) \le n-1 \tag{3.3.3}$$

there are k with $\operatorname{rank}(\Pi(\mathbf{A}) \cap \mathbb{Z}^n) \le k \le n-1$ and a tuple of positive reals $\mathbf{B} = (B_1, \ldots, B_N)$ with $N := \binom{n}{k}$ such that

(i) $B_1 \cdots B_N < Q(\mathbf{B})^{-\epsilon/2n^3}$;

(ii) $\operatorname{rank}(\Pi_k(\mathbf{B}) \cap \mathbb{Z}^N) = N - 1$;

(iii) $\Pi(\mathbf{A}) \cap \mathbb{Z}^n$ *is contained in a k-dimensional linear subspace V of $\mathbb{R}^n$ such that $f_k(V)$ is the $\mathbb{R}$-vector space generated by $\Pi_k(\mathbf{B}) \cap \mathbb{Z}^N$.*

Now assume that Theorem 3.2.1 has been proved for tuples $\mathbf{A}$ with $\operatorname{rank}(\Pi(\mathbf{A}) \cap \mathbb{Z}^n) = n-1$. Apply this to $\Pi_k(\mathbf{B})$ where k and $\mathbf{B}$ are determined according to Proposition 3.3.2. It follows that there are finitely many $(N-1)$-dimensional subspaces $S_1^{(k)}, \ldots, S_{t_k}^{(k)}$ of $\mathbb{R}^N$ such that for every $\mathbf{B}$ satisfying (i), (ii), the set $\Pi_k(\mathbf{B}) \cap \mathbb{Z}^N$ is contained in one of the spaces $S_j^{(k)}$ $(j = 1, \ldots, t_k)$. From (iii) and the fact that f_k is injective, we infer that for every tuple $\mathbf{A}$ with (3.3.3), the set $\Pi(\mathbf{A}) \cap \mathbb{Z}^n$ is contained in one of the spaces $T_j^{(k)}$ $(k = 1, \ldots, n-1,\ j = 1, \ldots, t_k)$, where $T_j^{(k)}$ is the unique k-dimensional subspace of $\mathbb{R}^n$ with $f_k(T_j^{(k)}) = S_j^{(k)}$. This implies Theorem 3.2.1 in full generality.

Proposition 3.3.2 can be proved by combining arguments from [65], Chap. IV, §§1, 3, 6, 7, Chap. VI, §§14, 15. Our approach is somewhat different. We need two lemmas.

3.3.4 Lemma. *(Minkowski's Second Theorem). Let C be a convex body in $\mathbb{R}^n$ which is symmetric about $\mathbf{0}$. For $\lambda > 0$, put $\lambda C := \{\lambda \mathbf{x} : \mathbf{x} \in C\}$. Define the n successive minima $\lambda_1, \ldots, \lambda_n$ of C by*

$$\lambda_i = \inf\{\lambda > 0 : \operatorname{rank}(\lambda C \cap \mathbb{Z}^n) = i\}.$$

Then

$$\frac{2^n}{n!} \le \lambda_1 \cdots \lambda_n \cdot \operatorname{vol}(C) \le 2^n.$$

Proof. E.g. [11], Lecture IV. □

The next lemma is to replace Davenport's lemma and Mahler's results on compound convex bodies used by Schmidt ([65], Chap. IV, §§3, 7).

3.3.5 Lemma. *Let W be an $\mathbb{R}$-vector space with basis $\{\mathbf{b}_1, \ldots, \mathbf{b}_n\}$, and let $l_1, \ldots, l_n$ be linearly independent linear functions from W to $\mathbb{R}$. Suppose that*

$$|l_i(\mathbf{b}_j)| \le \mu_j \qquad \text{for } i = 1, \ldots, n,\ j = 1, \ldots, n,$$

where $\mu_1 \le \cdots \le \mu_n$. Then there are a permutation κ of $\{1, \ldots, n\}$ and vectors

$$\begin{aligned} \mathbf{v}_1 &= \mathbf{b}_1 \\ \mathbf{v}_2 &= \mathbf{b}_2 + \xi_{21}\mathbf{b}_1 \\ &\vdots \\ \mathbf{v}_n &= \mathbf{b}_n + \xi_{n1}\mathbf{b}_1 + \cdots + \xi_{n,n-1}\mathbf{b}_{n-1} \end{aligned}$$

with $\xi_{ij} \in \mathbb{Z}$ for $1 \le j < i \le n$, such that

$$|l_{\kappa(i)}(\mathbf{v}_j)| \le 2^{i+j}\mu_{\min(i,j)} \qquad \text{for } i = 1, \ldots, n,\ j = 1, \ldots, n. \tag{3.3.6}$$

Proof. We proceed by induction on n. For $n = 1$, Lemma 3.3.5 is trivial. Suppose that Lemma 3.3.5 holds true for $n-1$ instead of n, where $n \geq 2$. Let W' denote the space generated by $\mathbf{b}_1, \ldots, \mathbf{b}_{n-1}$. There are $\alpha_1, \ldots, \alpha_n \in \mathbb{R}$, not all zero, such that

$$\alpha_1 l_1(\mathbf{x}) + \cdots + \alpha_n l_n(\mathbf{x}) = 0 \qquad \text{for } \mathbf{x} \in W'.$$

Choose $\kappa(n) \in \{1, \ldots, n\}$ such that $|\alpha_{\kappa(n)}| = \max(|\alpha_1|, \ldots, |\alpha_n|)$. Thus, with $\beta_j = -\alpha_j/\alpha_{\kappa(n)}$, we have

$$(3.3.7) \quad l_{\kappa(n)}(\mathbf{x}) = \sum_{j \neq \kappa(n)} \beta_j l_j(\mathbf{x}) \qquad \text{for } \mathbf{x} \in W', \text{ where } |\beta_j| \leq 1 \text{ for } j \neq \kappa(n).$$

By the induction hypothesis, there are a permutation $\kappa(1), \ldots, \kappa(n-1)$ of the set $\{1, \ldots, n\} \backslash \{\kappa(n)\}$, and vectors $\mathbf{v}_1 = \mathbf{b}_1, \ldots, \mathbf{v}_{n-1} = \mathbf{b}_{n-1} + \xi_{n-1,1}\mathbf{b}_1 + \cdots + \xi_{n-1,n-2}\mathbf{b}_{n-2}$ with $\xi_{ij} \in \mathbb{Z}$ for $1 \leq j < i \leq n-1$, such that

$$|l_{\kappa(i)}(\mathbf{v}_j)| \leq 2^{i+j}\mu_{\min(i,j)} \qquad \text{for } i = 1, \ldots, n-1,\ j = 1, \ldots, n-1.$$

By (3.3.7) we have $|l_{\kappa(n)}(\mathbf{v}_j)| \leq \sum_{i=1}^{n-1} |l_{\kappa(i)}(\mathbf{v}_j)| \leq (2^{1+j} + 2^{2+j} + \cdots + 2^{n-1+j})\mu_j < 2^{n+j}\mu_j$ for $j = 1, \ldots, n-1$. Hence

$$(3.3.8) \qquad |l_{\kappa(i)}(\mathbf{v}_j)| \leq 2^{i+j}\mu_{\min(i,j)} \qquad \text{for } i = 1, \ldots, n,\ j = 1, \ldots, n-1.$$

It suffices to show that there are $\xi_1, \ldots, \xi_{n-1} \in \mathbb{Z}$ such that $\mathbf{v}_n := \mathbf{b}_n + \xi_1\mathbf{v}_1 + \cdots + \xi_{n-1}\mathbf{v}_{n-1}$ satisfies

$$(3.3.9) \qquad |l_{\kappa(i)}(\mathbf{v}_n)| \leq 2^{i+n}\mu_i \qquad \text{for } i = 1, \ldots, n.$$

Define the vectors

$$\begin{aligned} \mathbf{a}_j &= (l_{\kappa(1)}(\mathbf{v}_j), \ldots, l_{\kappa(n)}(\mathbf{v}_j)) \qquad (j = 1, \ldots, n-1), \\ \mathbf{b} &= (l_{\kappa(1)}(\mathbf{b}_n), \ldots, l_{\kappa(n)}(\mathbf{b}_n)). \end{aligned}$$

Since $\{l_1, \ldots, l_n\}$ is linearly independent and $\{\mathbf{v}_1, \ldots, \mathbf{v}_{n-1}\}$ is a basis of W', and by (3.3.7), the set $\{\mathbf{a}_1, \ldots, \mathbf{a}_{n-1}\}$ is a basis of the space $x_n = \sum_{j=1}^{n-1} \beta_{\kappa(j)} x_j$. Hence there are $t_1, \ldots, t_{n-1} \in \mathbb{R}$ such that

$$\mathbf{b} = t_1\mathbf{a}_1 + \cdots + t_{n1}\mathbf{a}_{n-1} + (0, \ldots, 0, \alpha), \qquad \text{with } \alpha = l_{\kappa(n)}(\mathbf{b}_n) - \textstyle\sum_{j=1}^{n-1} \beta_{\kappa(j)} l_{\kappa(j)}(\mathbf{b}_n).$$

Choose $\xi_1, \ldots, \xi_{n-1} \in \mathbb{Z}$ such that $|\xi_j + t_j| \leq \frac{1}{2}$ for $j = 1, \ldots, n-1$. Put $\mathbf{v}_n := \mathbf{b}_n + \xi_1\mathbf{v}_1 + \cdots + \xi_{n-1}\mathbf{v}_{n-1}$. Thus,

$$(l_{\kappa(1)}(\mathbf{v}_n), \ldots, l_{\kappa(n)}(\mathbf{v}_n)) = \delta_1\mathbf{a}_1 + \cdots + \delta_{n-1}\mathbf{a}_{n-1} + (0, \ldots, 0, \alpha)$$

with $|\delta_j| \leq \frac{1}{2}$ for $j = 1, \ldots, n-1$. Put $\alpha_i = 0$ for $i = 1, \ldots, n-1$ and $\alpha_n = \alpha$. For $i = 1, \ldots, n$ we have, by assumption, $|l_{\kappa(i)}(\mathbf{b}_n)| \leq \mu_n$. Further, $|\beta_i| \leq 1$ for $i \neq \kappa(n)$. Hence $|\alpha| \leq n\mu_n$. Therefore, $|\alpha_i| \leq i\mu_i$ for $i = 1, \ldots, n$. Together with (3.3.8) this implies for $i = 1, \ldots, n$,

$$\begin{aligned} |l_{\kappa(i)}(\mathbf{v}_n)| &= |\delta_1 l_{\kappa(i)}(\mathbf{v}_1) + \cdots + \delta_{n-1} l_{\kappa(i)}(\mathbf{v}_{n-1}) + \alpha_i| \\ &\leq (|l_{\kappa(i)}(\mathbf{v}_1)| + \cdots + |l_{\kappa(i)}(\mathbf{v}_{n-1})|)/2 + |\alpha_i| \\ &\leq \{(2^{i+1} + 2^{i+2} + \cdots + 2^{i+n-1})/2 + i\}\mu_i \leq 2^{i+n}\mu_i. \end{aligned}$$

Hence (3.3.9) is satisfied. This completes the proof of Lemma 3.3.5. □

Proof. (of Proposition 3.3.2) Let $\mathbf{A} = (A_1, \dots, A_n)$ be a tuple satisfying (3.3.3). The parallelepiped $\Pi(\mathbf{A})$ is convex and symmetric about $\mathbf{0}$. Let $\lambda_1, \dots, \lambda_n$ be its successive minima. The volume of $\Pi(\mathbf{A})$ is $|\det(l_1, \dots, l_n)|^{-1} A_1 \cdots A_n$. Hence by Lemma 3.3.4,

$$\lambda_1 \cdots \lambda_n A_1 \cdots A_n \ll\gg 1. \tag{3.3.10}$$

Put $r := \mathrm{rank}(\Pi(\mathbf{A}) \cap \mathbb{Z}^n)$. Then $\lambda_r \le 1 < \lambda_{r+1}$. Further, by (3.3.10) and (3.3.3),

$$\lambda_n \ge (\lambda_1 \cdots \lambda_n)^{1/n} \gg (A_1 \cdots A_n)^{-1/n} > Q(\mathbf{A})^{\epsilon/n}.$$

The integer k in Proposition 3.3.2 is chosen from $\{r, r+1, \dots, n-1\}$ such that the quotient λ_k/λ_{k+1} is minimal. Thus,

$$\frac{\lambda_k}{\lambda_{k+1}} \le \left(\frac{\lambda_r}{\lambda_{r+1}} \cdot \frac{\lambda_{r+1}}{\lambda_{r+2}} \cdots \frac{\lambda_{n-1}}{\lambda_n} \right)^{1/(n-r)} \le \left(\frac{\lambda_r}{\lambda_n} \right)^{1/(n-1)}.$$

Hence

$$\frac{\lambda_k}{\lambda_{k+1}} \ll Q(\mathbf{A})^{-\epsilon/n(n-1)}. \tag{3.3.11}$$

Let $\mathbf{g}_1, \dots, \mathbf{g}_n$ be linearly independent vectors such that

$$\mathbf{g}_j \in \lambda_j \Pi(\mathbf{A}) \cap \mathbb{Z}^n \qquad \text{for } j = 1, \dots, n.$$

For $j = 1, \dots, N$, put

$$\hat{A}_j = A_{i_1} \cdots A_{i_{n-k}}, \quad \hat{\lambda}_j = \lambda_{i_1} \cdots \lambda_{i_{n-k}}, \quad \mathbf{G}_j = \mathbf{g}_{i_1} \wedge \cdots \wedge \mathbf{g}_{i_{n-k}},$$

where $\sigma_j = \{i_1 < \cdots < i_{n-k}\}$. Let V be the vector space generated by $\mathbf{g}_1, \dots, \mathbf{g}_k$. We know that $\lambda_r \le 1 < \lambda_{r+1}$. Hence the space generated by $\Pi(\mathbf{A}) \cap \mathbb{Z}^n$ is the same as that generated by $\mathbf{g}_1, \dots, \mathbf{g}_r$. Further, $r \le k$. Hence $\Pi(\mathbf{A}) \cap \mathbb{Z}^n \subset V$. Denote by W the space generated by $\mathbf{G}_1, \dots, \mathbf{G}_{N-1}$. Thus,

$$\Pi(\mathbf{A}) \cap \mathbb{Z}^n \subset V, \qquad W = f_k(V). \tag{3.3.12}$$

Laplace's rule states that for any vectors $\mathbf{a}_1, \dots, \mathbf{a}_{n-k}, \mathbf{b}_1, \dots, \mathbf{b}_{n-k} \in \mathbb{R}^n$,

$$(\mathbf{a}_1 \wedge \cdots \wedge \mathbf{a}_{n-k}, \mathbf{b}_1 \wedge \cdots \wedge \mathbf{b}_{n-k}) = \det((\mathbf{a}_i, \mathbf{b}_j)_{1 \le i,j \le n-k}).$$

In particular,

$$L_p(\mathbf{G}_q) = \det(l_r(\mathbf{g}_s))_{r \in \sigma_p, s \in \sigma_q}.$$

Let $\sigma_p = \{i_1 < \cdots < i_{n-k}\}$, $\sigma_q = \{j_1 < \cdots < j_{n-k}\}$. A typical summand of this determinant is $a(\tau) = \pm l_{i_1}(\mathbf{g}_{\tau(j_1)}) \cdots l_{i_{n-k}}(\mathbf{g}_{\tau(j_{n-k})})$, where τ is a permutation of $(j_1, \dots, j_{n-k})$. Note that

$$|a(\tau)| \le A_{i_1} \lambda_{\tau(j_1)} \cdots A_{i_{n-k}} \lambda_{\tau(j_{n-k})} = \hat{A}_p \hat{\lambda}_q.$$

Hence

$$|L_p(\mathbf{G}_q)| \le (n-k)! \hat{A}_p \hat{\lambda}_q \qquad \text{for } p = 1, \dots, N,\ q = 1, \dots, N.$$

We apply Lemma 3.3.5 to the forms $\hat{A}_1^{-1} L_1, \dots, \hat{A}_N^{-1} L_N$ and the vectors $\mathbf{G}_1, \dots, \mathbf{G}_N$. It follows that there are a permutation κ of $\{1, \dots, N\}$ and a basis $\{\mathbf{V}_1, \dots, \mathbf{V}_{N-1}\}$ of the space W generated by $\{\mathbf{G}_1, \dots, \mathbf{G}_{N-1}\}$, with $\mathbf{V}_1, \dots, \mathbf{V}_{N-1} \in \mathbb{Z}^N$, such that

$$\hat{A}_{\kappa(p)}^{-1} |L_{\kappa(p)}(\mathbf{V}_q)| \le 2^{2N} (n-k)! \hat{\lambda}_{\min(p,q)} \qquad \text{for } p = 1, \dots, N,\ q = 1, \dots, N-1.$$

Choose $\mathbf{B} = (B_1, \ldots, B_N)$ such that

$$(3.3.13)\qquad \begin{aligned} B_{\kappa(p)} &= 2^{2N}(n-k)!\hat{\lambda}_p \hat{A}_{\kappa(p)} \qquad (p = 1, \ldots, N-1), \\ B_{\kappa(N)} &= 2^{2N}(n-k)!\hat{\lambda}_{N-1}\hat{A}_{\kappa(N)}. \end{aligned}$$

Thus, $|L_p(\mathbf{V}_q)| \leq B_p$ for $p = 1, \ldots N$, $q = 1, \ldots, N-1$, i.e.,

$$(3.3.14)\qquad \mathbf{V}_1, \ldots, \mathbf{V}_{N-1} \in \Pi_k(\mathbf{B}) \cap \mathbb{Z}^N.$$

We show that $\mathbf{B}$ satisfies (ii), (iii) and (i).

Proof. (of (ii)) Let the succcessive minima of $\Pi_k(\mathbf{B})$ be $\nu_1, \ldots, \nu_N$. The volume of $\Pi_k(\mathbf{B})$ is $|\det(L_1, \ldots, L_N)|^{-1} B_1 \cdots B_N$. Together with Lemma 3.3.4 this implies that $\nu_1 \cdots \nu_N \gg (B_1 \cdots B_N)^{-1}$. Equation (3.3.14) implies that $\nu_{N-1} \leq 1$. Hence

$$(3.3.15)\qquad \nu_N \gg (B_1 \cdots B_N)^{-1}.$$

Note that $\hat{A}_1 \cdots \hat{A}_N = (A_1 \cdots A_n)^{\binom{n-1}{k}}$, that $\hat{\lambda}_1 \cdots \hat{\lambda}_N = (\lambda_1 \cdots \lambda_n)^{\binom{n-1}{k}}$, and that $\sigma_{N-1} = \{k, k+2, \ldots, n\}, \sigma_N = \{k+1, k+2, \ldots, n\}$. Hence

$$\frac{\hat{\lambda}_{N-1}}{\hat{\lambda}_N} = \frac{\lambda_k \lambda_{k+2} \cdots \lambda_n}{\lambda_{k+1}\lambda_{k+2} \cdots \lambda_n} = \frac{\lambda_k}{\lambda_{k+1}}.$$

Together with (3.3.13), (3.3.10) and (3.3.11), this gives

$$(3.3.16)\qquad \begin{aligned} B_1 \cdots B_N &\ll \hat{\lambda}_1 \cdots \hat{\lambda}_N \hat{A}_1 \cdots \hat{A}_N \hat{\lambda}_{N-1}/\hat{\lambda}_N = \\ &= (\lambda_1 \cdots \lambda_n A_1 \cdots A_n)^{\binom{n-1}{k}} \lambda_k/\lambda_{k+1} \ll Q(\mathbf{A})^{-\epsilon/n(n-1)}. \end{aligned}$$

Together with (3.3.15) this implies that $\nu_N > 1$ if $Q(\mathbf{A})$ is sufficiently large. Hence $\mathrm{rank}(\Pi_k(\mathbf{B}) \cap \mathbb{Z}^N) = N - 1$. □

Proof. (of (iii)) From (3.3.14), the fact that $\{\mathbf{V}_1, \ldots, \mathbf{V}_{N-1}\}$ is a basis of W and from $\mathrm{rank}(\Pi_k(\mathbf{B}) \cap \mathbb{Z}^N) = N-1$, it follows that the space generated by $\Pi_k(\mathbf{B}) \cap \mathbb{Z}^N$ is equal to W. Together with (3.3.12) this proves (iii). □

Proof. (of (i)) We estimate $Q(\mathbf{B})$ from above in terms of $Q(\mathbf{A})$ and insert this into (3.3.16). Note that $\mathbf{g}_1 \in \lambda_1 \Pi(\mathbf{A}) \cap \mathbb{Z}^n$ and $\mathbf{g}_1 \neq 0$. Hence

$$1 \leq |\mathbf{g}_1| \ll \max(|l_1(\mathbf{g}_1)|, \ldots, |l_n(\mathbf{g}_1)|) \leq \lambda_1 Q(\mathbf{A}).$$

Therefore, $\lambda_1 \gg Q(\mathbf{A})^{-1}$. Since each $\hat{\lambda}_j$ is the product of $n-k$ λ_i's, and each λ_i is bounded below by λ_1, this implies that

$$\hat{\lambda}_j \gg Q(\mathbf{A})^{-(n-k)} \qquad \text{for } j = 1, \ldots, N.$$

On the other hand, by (3.3.10),

$$\hat{\lambda}_j \leq (\lambda_1 \cdots \lambda_n)\lambda_1^{-k} \ll (A_1 \cdots A_n)^{-1} Q(\mathbf{A})^k \leq Q(\mathbf{A})^{n+k}.$$

By combining this with (3.3.13), and using that each $\hat{A}_j$ is the product of $n-k$ A_i's we get

$$\begin{aligned} Q(\mathbf{B}) &\ll \max(\hat{\lambda}_1, \ldots, \hat{\lambda}_N, \hat{\lambda}_1^{-1}, \ldots, \hat{\lambda}_N^{-1}) \max(\hat{A}_1, \ldots, \hat{A}_N, \hat{A}_1^{-1}, \ldots, \hat{A}_N^{-1}) \\ &\ll Q(\mathbf{A})^{(n+k)+(n-k)} = Q(\mathbf{A})^{2n}. \end{aligned}$$

Together with (3.3.16) and $Q(\mathbf{A})$ sufficiently large this implies that

$$B_1 \cdots B_N < Q(\mathbf{B})^{-\epsilon/2n^3},$$

as required. □

□

3.4 Step 2 of the Proof of Theorem 3.2.1

For this subsection we shall refer to [65], Chapters V, VI for details; $l_1, \ldots, l_n$ will be linearly independent linear forms in n variables with real algebraic integer coefficients, and $0 < \epsilon < 1$.

A generalisation of Roth's lemma.

Let $n \geq 2$, $m \geq 2$. For $h = 1, \ldots, m$, denote by $\mathbf{X}_h$ the block of n variables $(X_{h1}, \ldots, X_{hn})$. For a tuple $\mathbf{d} = (d_1, \ldots, d_m)$ of positive integers, denote by $\mathcal{R}(\mathbf{d})$ the set of polynomials in mn variables $P(\mathbf{X}_1, \ldots, \mathbf{X}_m) \in \mathbb{Z}[\mathbf{X}_1, \ldots, \mathbf{X}_m]$ such that P is homogeneous of degree d_h in the block of variables $\mathbf{X}_h$. We shall consider polynomials $P \in \mathcal{R}(\mathbf{d})$ as functions on the m-fold cartesian product $\overline{\mathbb{Q}}^n \times \cdots \times \overline{\mathbb{Q}}^n$. For a tuple $\mathbf{i} = (i_{11}, \ldots, i_{mn}) \in (\mathbb{Z}_{\geq 0})^{mn}$, put

$$P_{\mathbf{i}}(\mathbf{X}_1, \ldots, \mathbf{X}_m) = \frac{1}{i_{11}! \cdots i_{mn}!} \left(\frac{\partial}{\partial X_{11}}\right)^{i_{11}} \cdots \left(\frac{\partial}{\partial X_{mn}}\right)^{i_{mn}} P.$$

Note that for $P \in \mathcal{R}(\mathbf{d})$, the coefficients of $P_{\mathbf{i}}$ are integers. The factorials have been included to keep the coefficients of $P_{\mathbf{i}}$ small: if $H(P)$ denotes the height (maximum of absolute values of coefficients) of P, then

$$H(P_{\mathbf{i}}) \leq 2^{d_1 + \cdots + d_m} H(P).$$

For a tuple $\mathbf{i}$ as above, put

$$(\mathbf{i}/\mathbf{d}) = \sum_{h=1}^{m} \frac{i_{h1} + \cdots + i_{hn}}{d_h}.$$

3.4.1 Definition. Let $\mathbf{x} = (\mathbf{x}_1, \ldots, \mathbf{x}_m) \in \overline{\mathbb{Q}}^n \times \cdots \times \overline{\mathbb{Q}}^n$ (m times). The *index* of $\mathbf{x}$ w.r.t. P, denoted by $i(\mathbf{x}, P)$, is the largest number σ such that $P_{\mathbf{i}}(\mathbf{x}) = 0$ for all $\mathbf{i}$ with $(\mathbf{i}/\mathbf{d}) < \sigma$. Clearly,$\sigma \in \mathbb{Q}$. □

Every linear subspace V of $\mathbb{Q}^n$ of dimension $n-1$ can be given by an equation $a_1 X_1 + \cdots + a_n X_n = 0$ with $a_1, \ldots, a_n \in \mathbb{Z}$, $\gcd(a_1, \ldots, a_n) = 1$ which is uniquely determined up to sign. The *height* of V is defined by

$$H(V) := \max(|a_1|, \ldots, |a_n|).$$

3.4.2 Proposition. *(Generalisation of Roth's lemma). Let $0 < \sigma < 1$, $0 < \gamma \leq n-1$, m a positive integer, $d_1, \ldots, d_m$ positive integers, $V_1, \ldots, V_m$ $(n-1)$-dimensional linear subspaces of $\mathbb{Q}^n$, and P a polynomial with*

$$\begin{aligned}
\frac{d_h}{d_{h+1}} &\geq C_1 && \text{for } h = 1, \ldots, m-1; \\
H(V_h)^{d_h} &\geq H(V_1)^{d_1 \gamma} && \text{for } h = 2, \ldots, m, \\
H(V_h) &\geq C_2^{\gamma^{-1}(n-1)^2} && \text{for } h = 1, \ldots, m, \\
P &\in \mathcal{R}(\mathbf{d}), && \\
H(P) &\leq H(V_1)^{C_3 \gamma (n-1)^{-2} d_1},
\end{aligned}$$

where C_1, C_2 and C_3 are positive numbers depending only on m and σ. Then there is a point $\mathbf{x} \in V_1 \times \cdots \times V_m$ with $i(\mathbf{x}, P) < m\sigma$.

Proof. ([65], pp. 191–194.) The idea is as follows. For $n = 2$, Proposition 3.4.2 is a homogeneous version of Roth's lemma, except for the additional parameter γ. But with this parameter, the proof of Roth's lemma does not essentially change, cf. [65], pp. 137–148. The general case is reduced to $n = 2$ as follows. We may assume by permuting the variables in each block $\mathbf{X}_h$ if necessary, that V_h is given by an equation $a_{h1}X_1 + \cdots + a_{hn}X_n = 0$ with $a_{h1}, \ldots, a_{hn} \in \mathbb{Z}$, $\gcd(a_{h1}, \ldots, a_{hn}) = 1$, $a_{h1} = \max(|a_{h1}|, \ldots, |a_{hn}|)$, and $g := \gcd(a_{h1}, a_{h2}) \leq a_{h1}^{(n-2)/(n-1)}$. Put $P^* = P(\mathbf{X}_1^*, \ldots, \mathbf{X}_m^*)$, where $\mathbf{X}_h^* = (X_{h1}, X_{h2}, 0, \ldots, 0)$ and $V_h^* = \{a_{h1}X_1 + a_{h2}X_2 = 0\}$. We have $H(V_h^*) = a_{h1}/g$ and therefore $H(V_h)^{1/(n-1)} \leq H(V_h^*) \leq H(V_h)$. Now the conditions of Proposition 3.4.2 are satisfied with $n = 2$, and P^*, $V_1^*, \ldots, V_m^*$, $\gamma^* := \gamma/(n-1)$ replacing P, $V_1, \ldots, V_m$, γ. Hence $i(\mathbf{x}^*, P^*) < m\sigma$, with $\mathbf{x}^* = (\mathbf{x}_1^*, \ldots, \mathbf{x}_m^*), \mathbf{x}_h^* = (a_{h2}, -a_{h1})$ for $h = 1, \ldots, m$. This implies that $i(\mathbf{x}, P) < m\sigma$ for $\mathbf{x} = (\mathbf{x}_1, \ldots, \mathbf{x}_m)$ with $\mathbf{x}_h = (a_{h2}, -a_{h1}, 0, \ldots, 0) \in V_h$ for $h = 1, \ldots, m$. □

Construction of the auxiliary polynomial.

Introduce new variables

$$U_{hj} = l_j(X_{h1}, \ldots, X_{hn}) \qquad (h = 1, \ldots, m, \, j = 1, \ldots, n).$$

Since the linear forms $l_1, \ldots, l_n$ are linearly independent, we can express every polynomial $P \in \mathcal{R}(\mathbf{d})$ as

$$P = \sum_{\mathbf{j}} d_P(\mathbf{j}) U_{11}^{j_{11}} U_{12}^{j_{12}} \cdots U_{mn}^{j_{mn}},$$

where $\mathbf{j} = (j_{11}, \ldots, j_{mn})$ runs through all tuples of non-negative integers with $j_{h1} + \cdots + j_{hn} = d_h$ for $h = 1, \ldots, m$. Similarly, we can express $P_{\mathbf{i}}$ as

$$P_{\mathbf{i}} = \sum_{\mathbf{j}} d_{P,\mathbf{i}}(\mathbf{j}) U_{11}^{j_{11}} \cdots U_{mn}^{j_{mn}}.$$

We need the following generalisation of Roth's Index theorem, which states that there is a $P \in \mathcal{R}(\mathbf{d})$ of small height such that if $(\mathbf{i}/\mathbf{d})$ is small then $d_{P,\mathbf{i}}(\mathbf{j}) = 0$ unless $j_{1i}/d_1 + \cdots + j_{mi}/d_m \approx m/n$ for $i = 1, \ldots, n$.

3.4.3 Proposition. *(Polynomial theorem). Let $\sigma > 0$ and $\mathbf{d} = (d_1, \ldots, d_m)$ be any tuple of positive integers with $m > C_4(n, l_1, \ldots, l_n, \sigma)$. Then there is a polynomial $P \in \mathcal{R}(\mathbf{d})$ with the following properties:*

(i) *$H(P) \leq C_5^{d_1+\cdots+d_m}$, and $|d_{P,\mathbf{i}}(\mathbf{j})| \leq C_5^{d_1+\cdots+d_m}$ for all pairs $\mathbf{i}$, $\mathbf{j}$, where $C_5 = C_5(n, l_1, \ldots, l_n, \sigma)$;*

(ii) *if $(\mathbf{i}/\mathbf{d}) < 2m\sigma$, then $d_{P,\mathbf{i}}(\mathbf{j}) = 0$ unless $|\sum_{h=1}^m j_{hi}/d_h - m/n| \leq 3nm\sigma$ for $i = 1, \ldots, n$.*

Proof. This is Theorem 7A of [65], p. 180. Its complete proof is on pp. 176–184 of [65]. The idea is as follows. First one shows the existence of a polynomial $P \in \mathcal{R}(\mathbf{d})$ such that

$$\begin{aligned} H(P) &\leq C_6(n, l_1, \ldots, l_n, \sigma)^{d_1+\cdots+d_m} \\ (3.4.4) \quad d_P(\mathbf{j}) &= 0 \quad \text{for all tuples } \mathbf{j} \text{ with} \\ &\quad \textstyle\sum_{h=1}^m j_{hi}/d_h < (1/n - \sigma)m \qquad \text{for at least one } i \in \{1, \ldots, n\} \end{aligned}$$

(This is the Index Theorem on p. 176 of [65]). The proof is similar to that of the Index Theorem of Roth: one counts the number of tuples $\mathbf{j}$ in (3.4.4) by using a combinatorial argument (cf. [65], p. 122, Lemma 4C) and then applies Siegel's Lemma to the equations $d_P(\mathbf{j}) = 0$.

It is straightforward to show that P satisfies (i). Further, from the definition of index it follows that if $(\mathbf{i}, \mathbf{d}) < 2m\sigma$ then $d_{P,\mathbf{i}}(\mathbf{j}) = 0$ unless $\sum_{h=1}^{m} j_{hi}/d_h \geq (1/n - \sigma)m - 2m\sigma = m/n - 3m\sigma$ for $i = 1, \ldots, n$. But for such $\mathbf{i}, \mathbf{j}$ we also have

$$\sum_{h=1}^{m} \frac{j_{hi}}{d_h} = \sum_{k=1}^{n} \sum_{h=1}^{m} \frac{j_{hk}}{d_h} - \sum_{k \neq i} \left(\sum_{h=1}^{m} \frac{j_{hk}}{d_h} \right) \leq m - (n-1)(m/n - 3m\sigma) < m/n + 3nm\sigma.$$

This implies (ii). □

Proof. (of Theorem 3.2.1) By $C_7, C_8, \ldots$, we denote positive numbers depending only on $n, l_1, \ldots, l_n, \epsilon$. For a given tuple $\mathbf{A} = (A_1, \ldots, A_n)$ with $\operatorname{rank}(\Pi(\mathbf{A}) \cap \mathbb{Z}^n) = n - 1$, let $V_{\mathbf{A}}$ denote the $\mathbb{Q}$-vector space generated by $\Pi(\mathbf{A}) \cap \mathbb{Z}^n$. We must compare $Q(\mathbf{A})$ with $H(V_{\mathbf{A}})$.

3.4.5 Lemma. *Let $\mathbf{A} = (A_1, \ldots, A_n)$ be a tuple of positive reals with*

(3.4.6) $\quad A_1 \cdots A_n < Q(\mathbf{A})^{-\epsilon}, \qquad \operatorname{rank}(\Pi(\mathbf{A}) \cap \mathbb{Z}^n) = n - 1, \qquad H(V_{\mathbf{A}}) \geq C_7,$

where C_7 is sufficiently large. Then $Q(\mathbf{A})^{C_8} \leq H(V_{\mathbf{A}}) \leq Q(\mathbf{A})^{C_9}$.

Proof. This is essentially Lemma 11A of [65], p. 195. Schmidt used reciprocal parallelepipeds to prove this; we give a more straightforward proof. Suppose that $l_i(\mathbf{X}) = (\mathbf{c}_i, \mathbf{X})$ $(i = 1, \ldots, n)$, and that the coordinates of $\mathbf{c}_1, \ldots, \mathbf{c}_n$ are contained in an algebraic number field of degree D. Let $\mathbf{g}_1, \ldots, \mathbf{g}_{n-1}$ be linearly independent points in $\Pi(\mathbf{A}) \cap \mathbb{Z}^n$. For $k = 1, \ldots, n$, put

$$\Delta_k := \det((\mathbf{c}_i, \mathbf{g}_j))_{1 \leq i \leq n,\ i \neq k,\ 1 \leq j \leq n-1}.$$

Using that $|(\mathbf{c}_i, \mathbf{g}_j)| \leq A_i$ for $i = 1, \ldots, n$, we get

$$|\Delta_k| \ll A_1 \cdots A_{k-1} A_{k+1} \cdots A_n \ll Q(\mathbf{A})^{-\epsilon} A_k^{-1}. \tag{3.4.7}$$

We have $\mathbf{g}_1 \wedge \cdots \wedge \mathbf{g}_{n-1} = \lambda \mathbf{a}$, with $\lambda \in \mathbb{Z}$ and $\mathbf{a}$ being a vector with coprime integer coordinates. Further, $V_{\mathbf{A}}$ has equation $(\mathbf{a}, \mathbf{X}) = 0$. Therefore, $H(V_{\mathbf{A}}) = |\mathbf{a}|$. Put $\mathbf{c}_k^* := \mathbf{c}_1 \wedge \cdots \wedge \mathbf{c}_{k-1} \wedge \mathbf{c}_{k+1} \wedge \cdots \wedge \mathbf{c}_n$. Thus, by Laplace's rule, $\Delta_k = (\mathbf{c}_k^*, \mathbf{g}_1 \wedge \cdots \wedge \mathbf{g}_{n-1})$. Therefore,

$$|\Delta_k| = |\lambda| \cdot |(\mathbf{c}_k^*, \mathbf{a})| \geq |(\mathbf{c}_k^*, \mathbf{a})| \qquad \text{for } k = 1, \ldots, n. \tag{3.4.8}$$

Since $\{\mathbf{c}_1^*, \ldots, \mathbf{c}_n^*\}$ is linearly independent we have by (3.4.8) and (3.4.7)

$$\begin{aligned} H(V_{\mathbf{A}}) &= |\mathbf{a}| \ll \max(|\Delta_1|, \ldots, |\Delta_n|) \\ &\ll Q(\mathbf{A})^{-\epsilon} \max(A_1^{-1}, \ldots, A_n^{-1}) \\ &\ll Q(\mathbf{A})^{1-\epsilon}. \end{aligned} \tag{3.4.9}$$

Let $\Sigma = \{i : 1 \leq i \leq n,\ A_i \geq Q(\mathbf{A})^{-\epsilon/2}\}$. First suppose that $\Sigma \neq \emptyset$ and $\Delta_k \neq 0$ for some $k \in \Sigma$. For this k we have by (3.4.8) that $(\mathbf{c}_k^*, \mathbf{a}) \neq 0$. By an argument similar to that in the derivation of Theorem 1.5 from Theorem 3.2.1, we have $|(\mathbf{c}_k^*, \mathbf{a})| \gg |\mathbf{a}|^{-D}$. Together with (3.4.8), (3.4.7) this implies that

$$H(V_{\mathbf{A}}) = |\mathbf{a}| \gg Q(\mathbf{A})^{\epsilon/2D}.$$

This proves Lemma 3.4.5, provided that $\Sigma \neq \emptyset$, and $\Delta_k \neq 0$ for some $k \in \Sigma$.

Assume that $\Sigma = \emptyset$ or that $\Delta_k = 0$ for all $k \in \Sigma$. If $\Sigma = \emptyset$ then choose arbitrary linearly independent $\mathbf{h}_1, \ldots, \mathbf{h}_{n-1}$ from $\mathbb{Z}^n$. If $\Sigma \neq \emptyset$ and $\Delta_k = 0$ for all $k \in \sum$, then there are linearly independent integer vectors $\mathbf{h}_1, \ldots, \mathbf{h}_{n-1}$ such that $\det((\mathbf{c}_i, \mathbf{h}_j))_{1\le i\le n,\ i\neq k,\ 1\le j\le n-1} = 0$ for each $k \in \Sigma$. We can choose $\mathbf{h}_1, \ldots, \mathbf{h}_{n-1}$ from a finite set independent of $\mathbf{A}$, since there are only finitely many possibilities for Σ. Now it follows from $H(V_{\mathbf{A}}) \geq C_7$, that there is an $\mathbf{x} \in \Pi(\mathbf{A}) \cap \mathbb{Z}^n$ outside the space generated by $\mathbf{h}_1, \ldots, \mathbf{h}_{n-1}$, that is, with $|\det(\mathbf{x}, \mathbf{h}_1, \ldots, \mathbf{h}_{n-1})| \neq 0$. Since this number is a rational integer we have $|\det(\mathbf{x}, \mathbf{h}_1, \ldots, \mathbf{h}_{n-1})| \geq 1$. On the other hand, we have

$$\begin{aligned}\det(\mathbf{x}, \mathbf{h}_1, \ldots, \mathbf{h}_{n-1}) &= (\det(l_1, \ldots, l_n))^{-1} \sum_{k=1}^{n} \pm \alpha_k l_k(\mathbf{x}) \\ &= (\det(l_1, \ldots, l_n))^{-1} \sum_{k\notin\Sigma} \pm \alpha_k l_k(\mathbf{x}),\end{aligned}$$

where $\alpha_k := \det((\mathbf{c}_i, \mathbf{h}_j))_{i\neq k}$. Further, $|\alpha_k| \ll 1$, $|l_k(\mathbf{x})| \leq A_k$ for $k \notin \Sigma$. Hence by (3.4.9),

$$1 \leq |\det(\mathbf{x}, \mathbf{h}_1, \ldots, \mathbf{h}_{n-1})| \ll \sum_{k\notin\Sigma} A_k \ll Q(\mathbf{A})^{-\epsilon/2} \ll H(V_{\mathbf{A}})^{-\epsilon/2}.$$

But this contradicts that $H(V_{\mathbf{A}}) \geq C_7$. Therefore, $\Sigma \neq \emptyset$ and $\Delta_k \neq 0$ for some $k \in \Sigma$. □

We need another simple non-vanishing result for polynomials.

3.4.10 Lemma. *Let $f(X_1, \ldots, X_r) \in \mathbb{C}[X_1, \ldots, X_r]$ be a non-zero polynomial of degree $\leq s_h$ in X_h for $h = 1, \ldots, r$ and let $B_1, \ldots, B_r$ be positive reals. Then there are rational integers $x_1, \ldots, x_r, i_1, \ldots, i_r$ with*

$$|x_h| \leq B_h, \qquad 0 \leq i_h \leq s_h/B_h \qquad \text{for } h = 1, \ldots, r,$$

$$\left(\frac{\partial}{\partial X_1}\right)^{i_1} \cdots \left(\frac{\partial}{\partial X_r}\right)^{i_r} f(x_1, \ldots, x_r) \neq 0.$$

Proof. This is a special case of Lemma 8A on p. 184 of [65], which is sufficient for our purposes. For $r = 1$, Lemma 3.4.10 follows from the fact that f cannot be divisible by the polynomial $\{X(X-1)(X+1)(X-2)(X+2)\cdots(X-a)(X+a)\}^b$, where $a = [B_1]$, $b = [s_1/B_1] + 1$, since the latter polynomial is of degree $> s_1$. It is straightforward to complete the proof of Lemma 3.4.10 by induction on r. □

We assume that Theorem 3.2.1 is false. Then, by Proposition 3.3.2, there is an $n \geq 2$ such that if $\mathbf{A}$ runs through the tuples $\mathbf{A} = (A_1, \ldots, A_n)$ of positive reals with

$$\text{(3.4.11)} \quad \operatorname{rank}(\Pi(\mathbf{A}) \cap \mathbb{Z}^n) = n-1, \quad A_1 \cdots A_n < Q(\mathbf{A})^{-\epsilon}, \quad H(V_{\mathbf{A}}) \geq C_7,$$

then $V_{\mathbf{A}}$ runs through infinitely many $(n-1)$-dimensional linear subspaces of $\mathbb{Q}^n$. In view of Lemma 3.4.5, this implies that the set of numbers $Q(\mathbf{A})$ for which $\mathbf{A}$ satisfies (3.4.11) is unbounded.

Note that $A_i = Q(\mathbf{A})^{\varphi_i}$ for $i = 1,\dots,n$, where $\varphi_1,\dots,\varphi_n$ are reals depending on $\mathbf{A}$ with $|\varphi_i| \le 1$ for $i = 1,\dots,n$, and $\varphi_1 + \cdots + \varphi_n < -\epsilon$. The set of real points $(\varphi_1,\dots,\varphi_n)$ satisfying these inequalities can be covered by finitely many cubes $\{(\varphi_1,\dots,\varphi_n) : c_i - \epsilon/2n \le \varphi_i \le c_i \text{ for } i = 1,\dots,n\}$, where

$$|c_i| \le 1 \qquad \text{for } i = 1,\dots,n, \qquad c_1 + \cdots + c_n < -\epsilon/2. \tag{3.4.12}$$

Hence there is a tuple $(c_1,\dots,c_n)$ with (3.4.12), such that the set of numbers $Q(\mathbf{A})$ for tuples $\mathbf{A}$ with

$$\begin{aligned} \operatorname{rank}(\Pi(\mathbf{A}) \cap \mathbb{Z}^n) &= n-1, \\ A_1 \cdots A_n &< Q(\mathbf{A})^{-\epsilon}, \\ A_i &\le Q(\mathbf{A})^{c_i} \qquad \text{for } i = 1,\dots,n, \\ H(V_{\mathbf{A}}) &\ge C_7 \end{aligned} \tag{3.4.13}$$

is unbounded.

We choose σ sufficiently small, depending only on n and ϵ, with $0 < \sigma < 1/n$. Let m be the smallest integer exceeding the number $C_4(n, l_1,\dots,l_n,\sigma)$ in Proposition 3.4.3. Choose $\mathbf{A}_1,\dots,\mathbf{A}_m$ with (3.4.13) such that for sufficiently large C_{10}, C_{11},

$$Q(\mathbf{A}_1) \ge C_{10}, \qquad Q(\mathbf{A}_{h+1}) \ge Q(\mathbf{A}_h)^{C_{11}} \qquad \text{for } h = 2,\dots,m. \tag{3.4.14}$$

Put $Q_h := Q(\mathbf{A}_h)$, $V_h := V_{\mathbf{A}_h}$ for $h = 1,\dots,m$. Similarly as in the end of the proof of Roth's theorem, we choose a tuple $\mathbf{d} = (d_1,\dots,d_m)$ of positive integers such that

$$d_1\sigma \log Q_1 \ge \log Q_m, \tag{3.4.15}$$

and

$$d_1 \log Q_1 \le d_h \log Q_h \le (1+\sigma) d_1 \log Q_1 \qquad \text{for } h = 2,\dots,m. \tag{3.4.16}$$

The latter is possible since $\sigma d_1 \log Q_1 \ge \log Q_h$ for $h = 2,\dots,m$. Let $P \in \mathcal{R}(\mathbf{d})$ be the polynomial from Proposition 3.4.3. Put $\gamma := C_8/C_9$, where C_8, C_9 are the constants from Lemma 3.4.5. Let C_1, C_2 and C_3 be the constants from Proposition 3.4.2 and C_5 the constant from Proposition 3.4.3. We verify that $\sigma, m, \gamma, d_1,\dots,d_m, V_1,\dots,V_m$ and P satisfy the conditions of Proposition 3.4.2 for suitable C_{10}, C_{11}: namely, for sufficiently large C_{10} and C_{11} we have by (3.4.16), (3.4.14), $H(V_h) \ge C_7$, Lemma 3.4.5 and the upper bound for $H(P)$ in Proposition 3.4.3 that

$$\begin{aligned} &\frac{d_h}{d_{h+1}} = \frac{d_h \log Q_h}{d_{h+1}\log Q_{h+1}} \cdot \frac{\log Q_{h+1}}{\log Q_h} \ge (1+\sigma)^{-1} C_{11} > C_1 \qquad \text{for } h = 1,\dots,m-1, \\ &H(V_h)^{d_h} \ge Q_h^{C_8 d_h} \ge Q_1^{C_8 d_1} = Q_1^{C_9 d_1 \gamma} \ge H(V_1)^{d_1\gamma} \qquad \text{for } h = 1,\dots,m-1, \\ &H(V_h) \ge Q_h^{C_8} \ge Q_1^{C_8} \ge C_{10}^{C_8} \ge C_2^{\gamma^{-1}(n-1)^2} \qquad \text{for } h = 1,\dots,m \end{aligned}$$

and

$$\begin{aligned} H(V_1)^{C_3\gamma(n-1)^{-2}d_1} &\ge Q_1^{C_3C_8\gamma(n-1)^{-2}d_1} \ge C_{10}^{C_3C_8\gamma(n-1)^{-2}d_1} \\ &> C_5^{md_1} > C_5^{d_1+\cdots+d_m} \ge H(P). \end{aligned}$$

Now Proposition 3.4.2 implies that there is a point $\mathbf{x} \in V_1 \times \cdots \times V_m$ with $i(\mathbf{x}, P) < m\sigma$.

What we actually want is a point $\mathbf{x} = (\mathbf{x}_1, \ldots, \mathbf{x}_m)$ with $\mathbf{x}_h \in \Pi(\mathbf{A}_h) \cap \mathbb{Z}^n$ for $h = 1, \ldots, m$ and small index w.r.t. P. There is such a point in a slightly larger set. Namely, choose a tuple $\mathbf{i}$ with $(\mathbf{i}/\mathbf{d}) < \sigma$ such that $P_\mathbf{i}$ does not vanish everywhere on $V_1 \times \cdots \times V_m$. Choose linearly independent vectors $\mathbf{g}_{h1}, \ldots, \mathbf{g}_{h,n-1}$ from $\Pi(\mathbf{A}_h) \cap \mathbb{Z}^n$ for $h = 1, \ldots, m$ and define the polynomial

$$f(Y_{11}, \ldots, Y_{m,n-1}) = P_\mathbf{i}\left(\sum_{k=1}^{n-1} Y_{1k}\mathbf{g}_{1k}, \ldots, \sum_{k=1}^{n-1} Y_{mk}\mathbf{g}_{mk}\right).$$

Then f is not zero. By Lemma 3.4.10, there are integers $y_{11}, \ldots, y_{m,n-1}$, and integers $k_{11}, \ldots, k_{m,n-1}$ with

$$|y_{hi}| \le n\sigma^{-1}, \quad 0 \le k_{hi} \le \sigma d_h/n \qquad \text{for } h = 1, \ldots, m,\ i = 1, \ldots, n-1,$$
$$f^*(\mathbf{y}) := \left(\frac{\partial}{\partial Y_{11}}\right)^{k_{11}} \cdots \left(\frac{\partial}{\partial Y_{m,n-1}}\right)^{k_{m,n-1}} f(y_{11}, \ldots, y_{m,n-1}) \ne 0.$$

Put $\mathbf{x}_h := \sum_{i=1}^{n-1} y_{hi}\mathbf{g}_{hi}$ $(h = 1, \ldots, m)$, $\mathbf{x} = (\mathbf{x}_1, \ldots, \mathbf{x}_m)$, $\lambda\mathbf{A} := (\lambda A_1, \ldots, \lambda A_n)$ for $\mathbf{A} = (A_1, \ldots, A_n)$. Note that $f^*(\mathbf{y})$ is a linear combination of terms $P_{\mathbf{i}+\mathbf{j}}(\mathbf{x})$, where $\mathbf{j} = (j_{11}, \ldots, j_{mn})$ with $\sum_{l=1}^{n} j_{hl} = \sum_{l=1}^{n-1} k_{hl}$ for $h = 1, \ldots, m$. Hence

$$\text{(3.4.17)} \qquad \begin{array}{l} \mathbf{x}_h \in \Pi(n\sigma^{-1}\mathbf{A}_h) \cap \mathbb{Z}^n \qquad \text{for } h = 1, \ldots m, \\ i(\mathbf{x}, P) < m\sigma + \sum_{h=1}^{m} \frac{k_{h1}+\cdots+k_{h,n-1}}{d_h} \le 2m\sigma. \end{array}$$

Hence there is a tuple $\mathbf{i}$ with $(\mathbf{i}/\mathbf{d}) < 2m\sigma$, $P_\mathbf{i}(\mathbf{x}) \ne 0$. Note that $P_\mathbf{i}(\mathbf{x}) \in \mathbb{Z}$. Below we shall show that $|P_\mathbf{i}(\mathbf{x})| < 1$. Thus, our initial assumption that Theorem 3.2.1 is false leads to a contradiction.

Put $u_{hi} = l_i(\mathbf{x}_h)$ for $h = 1, \ldots, m$, $i = 1, \ldots, n$. By Proposition 3.4.3 we have

$$\text{(3.4.18)} \quad |P_\mathbf{i}(\mathbf{x})| \le \sum_{\mathbf{j}}^{*} |d_{P,\mathbf{i}}(\mathbf{j})||u_{11}^{j_{11}} \cdots u_{mn}^{j_{mn}}| \le C_{12}^{d_1+\cdots+d_m} \max_{\mathbf{j}}^{*} |u_{11}^{j_{11}} \cdots u_{mn}^{j_{mn}}|,$$

where the sum, maximum, respectively, are taken over all tuples $\mathbf{j} = (j_{11}, \ldots, j_{mn})$ with $|\sum_{h=1}^{m} j_{hi}/d_h - m/n| \le 3nm\sigma$ for $i = 1, \ldots, n$. Fix such a tuple $\mathbf{j}$. By (3.4.17) and (3.4.13) we have

$$|u_{hi}| \le n\sigma^{-1}Q_h^{c_i} \qquad \text{for } h = 1, \ldots, m,\ i = 1, \ldots, n.$$

Hence

$$\text{(3.4.19)} \quad \log|u_{11}^{j_{11}} \cdots u_{mn}^{j_{mn}}| \le \sum_{i=1}^{n}\sum_{h=1}^{m} c_i\left(\frac{j_{hi}}{d_h}\right) d_h \log Q_h + (d_1+\cdots+d_m)\log\left(\frac{n}{\sigma}\right).$$

Further, by (3.4.16) and $\sigma < 1$ we have

$$\begin{aligned} \sum_{h=1}^{m}(j_{hi}/d_h)d_h \log Q_h &\le d_1 \log Q_1 \cdot (1+\sigma) \cdot \sum_{h=1}^{m}(j_{hi}/d_h) \\ &\le d_1 \log Q_1 \cdot (1+\sigma)(m/n + 3nm\sigma) \\ &\le mn^{-1} d_1 \log Q_1 \cdot (1 + 7n^2\sigma), \\ \sum_{h=1}^{m}(j_{hi}/d_h)d_h \log Q_h &\ge d_1 \log Q_1 \sum_{h=1}^{m}(j_{hi}/d_h) \\ &\ge d_1 \log Q_1 (m/n - 3nm\sigma) \ge mn^{-1} d_1 \log Q_1 \cdot (1 - 7n^2\sigma). \end{aligned}$$

By inserting this into (3.4.19), and using (3.4.12), we get

$$\log |u_{11}^{j_{11}} \cdots u_{mn}^{j_{mn}}| \leq (mn^{-1} d_1 \log Q_1) \cdot \delta + (d_1 + \cdots + d_m) \log(n\sigma^{-1}),$$

with

$$\delta = \sum_{i=1}^{n} c_i + 7n^2\sigma \sum_{i=1}^{n} |c_i| \leq -\epsilon/2 + 7n^3\sigma < -\epsilon/3,$$

if we choose $\sigma < \epsilon/42n^3$. Together with (3.4.18) this yields, for sufficiently large C_{10},

$$|P_1(\mathbf{x})| \leq C_{13}^{d_1+\cdots+d_m} Q_1^{-md_1\epsilon/3n} \leq (C_{13} Q_1^{-\epsilon/3n})^{md_1} < 1,$$

which is the contradiction we wanted. This completes the proof of Theorem 3.2.1. □

Chapter V

Heights on Abelian Varieties

by Johan Huisman

1 Height on Projective Space

We will follow to a great extent Chapter 6 of [14].

If K is a number field and v a finite place of K, that is, v corresponds to a prime ideal $\mathfrak{p}$ of the ring of integers of K, then we define a norm $\|\cdot\|_v$ on K by

$$\|x\|_v = \left(\frac{1}{N\mathfrak{p}}\right)^{\mathrm{ord}_{\mathfrak{p}}(x)},$$

where $N\mathfrak{p}$ is the absolute norm of $\mathfrak{p}$. If v is an infinite place, that is, v corresponds to an embedding σ of K in $\mathbb{R}$ or v corresponds to a conjugate pair $\{\sigma, \overline{\sigma}\}$ of embeddings of K in $\mathbb{C}$, then we define a norm $\|\cdot\|_v$ on K by

$$\|x\|_v = \begin{cases} |\sigma(x)|, & \text{if } v \text{ is real,} \\ |\sigma(x)|^2, & \text{if } v \text{ is complex.} \end{cases}$$

Clearly, for any place v of K, the homothety $y \mapsto xy$ transforms a Haar measure μ on the completion K_v of K at v into $\|x\|_v \cdot \mu$. Let M_K be the set of places of K, M_K^∞ the set of infinite places and M_K^f the set of finite places. Then we have the *product formula*

$$\prod_{v \in M_K^f} \|x\|_v \prod_{\sigma: K \to \mathbb{C}} |\sigma(x)| = \prod_{v \in M_K} \|x\|_v = 1,$$

for every $x \in K^*$. This can be easily seen as follows (cf. [87], Ch. IV, §4, Thm. 5). Let A be the ring of adèles of K. The product formula will follow if we prove that any Haar measure on A is invariant under the homothety $\lambda_x: y \mapsto xy$ of A, for any $x \in K^*$. Since A/K is a compact topological group and λ_x induces an isomorphism $\overline{\lambda_x}$ of A/K, any Haar measure on A/K is invariant under $\overline{\lambda_x}$. Moreover, since K is discrete and the restriction $\widehat{\lambda_x}$ of λ_x to K is an isomorphism of K as a topological group, any Haar measure on K is invariant under $\widehat{\lambda_x}$. Therefore, any Haar measure on A is invariant under λ_x.

If $P = (x_0 : \cdots : x_n)$ is in $\mathbb{P}^n(K)$ then we define the *height of P relative to K* by

$$H_K(P) = \prod_{v \in M_K} \max\{\|x_0\|_v, \ldots, \|x_n\|_v\}.$$

Observe that, by the product formula, this is well defined.

1.1 Example. If $K = \mathbb{Q}$ and $P \in \mathbb{P}^n(\mathbb{Q})$ then we may assume $P = (x_0 : \cdots : x_n)$ with $x_i \in \mathbb{Z}$ and $\gcd(x_0, \ldots, x_n) = 1$. Then

$$H_{\mathbb{Q}}(P) = \max\{|x_0|, \ldots, |x_n|\}.$$

□

It is clear that $H_K(P) \geq 1$, for every $P \in \mathbb{P}^n(K)$. If L is a finite extension of K and $P \in \mathbb{P}^n(K)$ then

$$H_L(P) = H_K(P)^{[L:K]}.$$

Hence, we can define the *absolute height* H on $\mathbb{P}^n(\overline{K})$ by

$$H(P) = H_L(P)^{\frac{1}{[L:\mathbb{Q}]}},$$

where L is some number field containing the coordinates of P. It will be convenient to define the *(logarithmic) height* h on $\mathbb{P}^n(\overline{K})$ by

$$h(P) = \log H(P). \tag{1.1.1}$$

1.2 Theorem (Northcott). *Let n, d and C be integers. Then*

$$\{P \in \mathbb{P}^n(\overline{K}) \mid H(P) \leq C \quad \text{and} \quad [K(P) : K] \leq d\}$$

is a finite set.

For a proof the reader is referred to [14], Chapter 6 or [70], §2.4.

2 Heights on Projective Varieties

We will define height functions on a projective algebraic variety V over a number field K, using morphisms from V into projective space. Suppose

$$f: V \longrightarrow \mathbb{P}^n$$

is a morphism of algebraic varieties over K. Then one defines the *(logarithmic) height on V relative to f* by

$$\begin{aligned} h_f: V(\overline{K}) &\longrightarrow \mathbb{R} \\ P &\longmapsto h(f(P)). \end{aligned}$$

Let us call real-valued functions h and h' on the set $V(\overline{K})$ *equivalent*, denoted by $h \sim h'$, if $|h - h'|$ is bounded on $V(\overline{K})$. It turns out that the height h_f depends only, up to equivalence, on the invertible sheaf $\mathcal{L} = f^*\mathcal{O}_{\mathbb{P}^n}(1)$.

2.1 Theorem. *Let V be a projective algebraic variety over K. If $f: V \to \mathbb{P}^n$ and $g: V \to \mathbb{P}^m$ are morphisms over K such that*

$$f^*\mathcal{O}_{\mathbb{P}^n}(1) \cong g^*\mathcal{O}_{\mathbb{P}^m}(1)$$

then h_f and h_g are equivalent.

Proof. Recall that a morphism $\phi: V \to \mathbb{P}^k$ is uniquely determined by (the isomorphism class of) the invertible sheaf $\mathcal{L} = \phi^*\mathcal{O}(1)$ and the global sections $s_i = \phi^* x_i \in \Gamma(V, \mathcal{L})$. Therefore, it suffices to prove the theorem in the case that $m \geq n$ and $f = \pi \circ g$, where π is the rational map

$$\begin{array}{rccc} \pi: \mathbb{P}^m & \cdots\longrightarrow & \mathbb{P}^n \\ (x_0: \cdots : x_m) & \longmapsto & (x_0: \cdots : x_n). \end{array}$$

Clearly, $h_g - h_f \geq 0$. To prove that $h_g - h_f$ is bounded from above, observe that

$$g(V) \cap \mathfrak{V}(X_0, \ldots, X_n)$$

is empty. Since $g(V)$ is closed, let $I \subseteq K[X_0, \ldots, X_m]$ be its defining homogeneous ideal. Then

$$\sqrt{I + (X_0, \ldots, X_n)} = (X_0, \ldots, X_m)$$

in $K[X_0, \ldots, X_m]$. Therefore, there exist a positive integer q and $F_{ij} \in K[X_0, \ldots, X_m]$ such that

$$X_{n+i}^q - \sum_{j=0}^{n} F_{ij} X_j \in I, \quad i = 0, \ldots, m-n.$$

We may assume F_{ij} to be homogeneous of degree $q-1$. Denote the coefficients of F_{ij} by a_{ijk}. If $L \subseteq \overline{K}$ is a finite extension of K and w is a place of L then we define

$$\varepsilon_w = \begin{cases} 0 & \text{if } w \text{ is finite,} \\ 1 & \text{if } w \text{ is real,} \\ 2 & \text{if } w \text{ is complex} \end{cases}$$

and put

$$c_w = (n+1)^{\varepsilon_w} \binom{q-1+m}{m}^{\varepsilon_w} \cdot \max \|a_{ijk}\|_w.$$

Choose $P \in g(V)(L)$, say $P = (x_o: \cdots : x_m)$ with $x_i \in L$. It is easy to see that

$$\|x_{n+i}\|_w^q \leq c_w \cdot \max_{j \leq m} \|x_j\|_w^{q-1} \cdot \max_{j \leq n} \|x_j\|_w,$$

for $i = 0, \ldots, m-n$. Put

$$c'_w = \max\{1, c_w^{\frac{1}{q}}\},$$

then

$$\max_{i \leq m} \|x_i\|_w \leq c'_w \cdot \max_{j \leq n} \|x_j\|_w.$$

In particular,

$$\begin{aligned} H_L(x_0: \cdots : x_m) &= \prod_{w \in M_L} \max_{i \leq m} \|x_i\|_w \\ &\leq \left(\prod_{w \in M_L} c'_w\right) \left(\prod_{w \in M_L} \max_{j \leq n} \|x_j\|_w\right) \\ &= \left(\prod_{v \in M_K} c'_v\right)^d H_L(x_0: \cdots : x_n), \end{aligned}$$

where $d = [L:K]$. Therefore

$$h(x_0:\cdots:x_m) \le h(x_0:\cdots:x_n) + c,$$

where $c = \frac{1}{[K:\mathbb{Q}]}\sum_{v\in M_K} c'_v$ which neither depends on P nor on L. Hence $h_g - h_f$ is bounded from above. □

As a consequence, we can define, up to equivalence, a height function $h_{\mathcal{L}}$ for every invertible sheaf $\mathcal{L}$ on V which is basepoint-free (i.e., generated by global sections). For, choose a morphism f over K from V into $\mathbb{P}^n$ such that

$$\mathcal{L} \cong f^*\mathcal{O}_{\mathbb{P}^n}(1).$$

(Such a morphism exists since $\mathcal{L}$ is basepoint-free.) Then, by Theorem 2.1,

$$h_{\mathcal{L}} = h_f$$

depends only, up to equivalence, on $\mathcal{L}$. More precisely, one defines $h_{\mathcal{L}}$ as the equivalence class of h_f. That is, if $\mathcal{H}(V(\overline{K}))$ is the group of equivalence classes of real-valued functions on $V(\overline{K})$, we have $h_{\mathcal{L}} \in \mathcal{H}(V(\overline{K}))$. However, often we will treat $h_{\mathcal{L}}$ as a real-valued function, keeping in mind that $h_{\mathcal{L}}$ is only defined up to equivalence. It is easy to prove that, for any basepoint-free invertible sheaves $\mathcal{L}$ and $\mathcal{M}$ on V,

$$h_{\mathcal{L}\otimes\mathcal{M}} \sim h_{\mathcal{L}} + h_{\mathcal{M}}. \tag{2.2}$$

As a consequence, for any invertible sheaf $\mathcal{L}$ on V, we can define, up to equivalence, a height function $h_{\mathcal{L}}$ by

$$h_{\mathcal{L}} = h_{\mathcal{L}_1} - h_{\mathcal{L}_2},$$

where $\mathcal{L}_1$ and $\mathcal{L}_2$ are basepoint-free invertible sheaves on V such that

$$\mathcal{L} \cong \mathcal{L}_1 \otimes \mathcal{L}_2^{-1}.$$

(Such sheaves always exist; see [27], p. 121.) By (2.2), this does not depend on $\mathcal{L}_1$, $\mathcal{L}_2$. Hence the following result due to A. Weil.

2.3 Theorem. *Let V be a projective algebraic variety over K. There exists a unique homomorphism*

$$h: \mathrm{Pic}V \longrightarrow \mathcal{H}(V(\overline{K}))$$

such that

(i) *if $V = \mathbb{P}^n$ then $h_{\mathcal{O}_{\mathbb{P}^n}(1)}$ is the usual height h on projective space,*

(ii) *if W is a projective algebraic variety over K and $f: V \to W$ is a morphism over K then*

$$h_{f^*\mathcal{L}} = h_{\mathcal{L}} \circ f,$$

for any $\mathcal{L} \in \mathrm{Pic}W$.

It is then easy to prove, using Theorem 1.2, the following finiteness theorem.

2.4 Theorem. *Let V be a projective algebraic variety over K. If $\mathcal{L}$ is an ample invertible sheaf on V then, for all real numbers C and d, the set*

$$\{P \in V(\overline{K}) \mid h_{\mathcal{L}}(P) \le C \quad \text{and} \quad [K(P):K] \le d\}$$

is a finite set.

Observe that it makes sense to call an element of $\mathcal{H}(V)$ bounded from below (or above).

2.5 Theorem. *Let V be a projective algebraic variety over K. If $\mathcal{L}$ is an invertible sheaf on V and s is a global section then $h_{\mathcal{L}}$ is bounded from below on the set*

$$\{P \in V(\overline{K}) \mid s(P) \neq 0\}.$$

Proof. Choose basepoint-free invertible sheaves $\mathcal{L}_1$ and $\mathcal{L}_2$ on V such that

$$\mathcal{L} \cong \mathcal{L}_1 \otimes \mathcal{L}_2^{-1}.$$

Let $s_0, \ldots, s_n$ be global sections of $\mathcal{L}_2$ that generate $\mathcal{L}_2$. Choose global sections $s_{n+1}, \ldots, s_m$ of $\mathcal{L}_1$ such that

$$s \otimes s_0, \ldots, s \otimes s_n, s_{n+1}, \ldots, s_m$$

generate $\mathcal{L}_1$. Then, whenever $P \in V(\overline{K})$ with $s(P) \neq 0$,

$$\begin{aligned} h_{\mathcal{L}_1}(P) &= h(s \otimes s_0(P), \ldots, s \otimes s_n(P), s_{n+1}(P), \ldots, s_m(P)) \\ &\geq h(s \otimes s_0(P), \ldots, s \otimes s_n(P)) \\ &= h(s_0(P), \ldots, s_n(P)) \\ &= h_{\mathcal{L}_2}(P). \end{aligned}$$

Therefore, $h_{\mathcal{L}}$ is bounded from below on the set of $P \in V(\overline{K})$ such that $s(P) \neq 0$. □

3 Heights on Abelian Varieties

We will need the Theorem of the Cube.

3.1 Theorem (Theorem of the Cube). *Let X_1, X_2, X_3 be complete algebraic varieties over the field K and let $P_i \in X_i(K)$. Then, an invertible sheaf $\mathcal{L}$ on $X_1 \times X_2 \times X_3$ is trivial whenever its restrictions to $\{P_1\} \times X_2 \times X_3$, $X_1 \times \{P_2\} \times X_3$ and $X_1 \times X_2 \times \{P_3\}$ are trivial.*

Proof. (After [52], Chapter II, §6.) Let us give a proof when $\mathrm{char}(K) = 0$, since this is the case we are interested in. Then it suffices to prove the theorem for $K = \mathbb{C}$. (Here one uses that the X_i, P_i and $\mathcal{L}$ are defined over a subfield K_0 of L which is of finite type over $\mathbb{Q}$ and that for $K \to L$ any field extension, X any quasi-compact separated K-scheme and $\mathcal{L}$ any invertible sheaf on X, $\mathcal{L}$ is trivial if and only if its pullback to X_L is trivial.)

Before we continue the proof let us recall the following definition. A contravariant functor F from the category of complete complex algebraic varieties with basepoints into the category of abelian groups is called of order $\leq n$ if for all complete complex algebraic varieties $X_0, \ldots, X_n$ with basepoints, the natural mapping

$$F(\prod_{i=0}^{n} X_i) \longrightarrow \prod_{j=0}^{n} F(\prod_{i \neq j} X_i)$$

is injective. As an example, the Theorem of the Cube states that the functor Pic is of order ≤ 2.

To finish the proof we switch to an analytic point of view. Let $\mathcal{O}_{X,h}$ denote the sheaf of analytic functions on X. According to the GAGA-principle,

$$\mathrm{Pic}X \cong H^1(X, \mathcal{O}^*_{X,h}),$$

for any complete complex algebraic variety X. The long exact sequence associated to

$$0 \longrightarrow \mathbb{Z} \longrightarrow \mathcal{O}_{X,h} \xrightarrow{\exp} \mathcal{O}^*_{X,h} \longrightarrow 0$$

implies the existence of an exact sequence

$$H^1(X, \mathcal{O}_{X,h}) \longrightarrow H^1(X, \mathcal{O}^*_{X,h}) \longrightarrow H^2(X, \mathbb{Z}).$$

Since both $H^1(\cdot, \mathcal{O}_{X,h})$ and $H^2(\cdot, \mathbb{Z})$ are functors of order ≤ 2, the functor $H^1(\cdot, \mathcal{O}^*_{X,h})$ is of order ≤ 2. This proves the theorem. □

3.2 Corollary. *Let X be an abelian variety over the field K and let $p_i \colon X^3 \to X$ be the projection on the ith factor. Let $p_{ij} = p_i + p_j$ and $p_{ijk} = p_i + p_j + p_k$. Then, for any invertible sheaf $\mathcal{L}$ on X, the invertible sheaf*

$$p_{123}^*\mathcal{L} \otimes p_{12}^*\mathcal{L}^{-1} \otimes p_{13}^*\mathcal{L}^{-1} \otimes p_{23}^*\mathcal{L}^{-1} \otimes p_1^*\mathcal{L} \otimes p_2^*\mathcal{L} \otimes p_3^*\mathcal{L}$$

on X^3 is trivial.

Proof. Taking restrictions of this sheaf to $O \times X \times X$, $X \times O \times X$ and $X \times X \times O$ yields a trivial sheaf. The conclusion follows from the Theorem of the Cube. □

Since this corollary expresses a relation between sheaves on X^3, we have immediately, by Theorem 2.3, the following fact about heights on abelian varieties.

3.3 Theorem. *If X is an abelian variety over a number field K then, for any invertible sheaf $\mathcal{L}$ on X,*

$$h_{\mathcal{L}}(P+Q+R) - h_{\mathcal{L}}(P+Q) - h_{\mathcal{L}}(P+R) - h_{\mathcal{L}}(Q+R) + h_{\mathcal{L}}(P) + h_{\mathcal{L}}(Q) + h_{\mathcal{L}}(R) \sim 0,$$

as functions on $X(\overline{K})^3$.

Let us denote for an abelian variety X over K the multiplication-by-n mapping from X into itself by $[n]$, for any integer n. Recall that an invertible sheaf $\mathcal{L}$ is called symmetric (resp. antisymmetric) whenever $[-1]^*\mathcal{L} \cong \mathcal{L}$ (resp. $[-1]^*\mathcal{L} \cong \mathcal{L}^{-1}$). As a consequence of Corollary 3.2, one can prove the following.

3.4 Corollary. *If X is an abelian variety over the field K and $\mathcal{L}$ is an invertible sheaf on X then*

$$[n]^*\mathcal{L} \cong \mathcal{L}^{(n^2+n)/2} \otimes [-1]^*\mathcal{L}^{(n^2-n)/2},$$

for any integer n. In particular,

$$[n]^*\mathcal{L} \cong \begin{cases} \mathcal{L}^{n^2}, & \text{if } \mathcal{L} \text{ is symmetric,} \\ \mathcal{L}^{n}, & \text{if } \mathcal{L} \text{ is antisymmetric.} \end{cases}$$

Again, this translates into properties of heights on abelian varieties.

3.5 Theorem. *If X is an abelian variety over the number field K and $\mathcal{L}$ is an invertible sheaf on X then*

$$h_{\mathcal{L}} \circ [n] \sim \tfrac{n^2+n}{2} h_{\mathcal{L}} + \tfrac{n^2-n}{2} h_{\mathcal{L}} \circ [-1],$$

for any integer n. In particular,

$$h_{\mathcal{L}} \circ [n] \sim \begin{cases} n^2 h_{\mathcal{L}}, & \text{if } \mathcal{L} \text{ is symmetric,} \\ n h_{\mathcal{L}}, & \text{if } \mathcal{L} \text{ is antisymmetric.} \end{cases}$$

The property of $h_{\mathcal{L}}$ in Theorem 3.3 will imply the existence of a canonical height function, the *Néron-Tate height relative to $\mathcal{L}$*,

$$\hat{h}_{\mathcal{L}}: X(\overline{K}) \longrightarrow \mathbb{R},$$

as stated in the following theorem.

3.6 Theorem. *If X is an abelian variety over the number field K and $\mathcal{L}$ is an invertible sheaf on X then there exist a unique symmetric bilinear mapping $b_{\mathcal{L}}: X(\overline{K}) \times X(\overline{K}) \to \mathbb{R}$ and a unique linear mapping $l_{\mathcal{L}}: X(\overline{K}) \to \mathbb{R}$, such that*

$$\hat{h}_{\mathcal{L}}: X(\overline{K}) \longrightarrow \mathbb{R},$$

defined by

$$\hat{h}_{\mathcal{L}}(P) = \tfrac{1}{2} b_{\mathcal{L}}(P,P) + l_{\mathcal{L}}(P),$$

is equivalent to $h_{\mathcal{L}}$. Moreover, if $\mathcal{L}$ is symmetric then $l_{\mathcal{L}} = 0$ and if $\mathcal{L}$ is ample then $b_{\mathcal{L}}$ is positive definite on $X(\overline{K}) \otimes \mathbb{R}$.

Proof. The existence and uniqueness of $b_{\mathcal{L}}$ and $l_{\mathcal{L}}$ follow from the lemma below, whose proof can be found in [52], Appendix II, §5.

If $\mathcal{L}$ is symmetric then in virtue of Theorem 3.5

$$\tfrac{1}{2} n^2 b_{\mathcal{L}}(P,P) + n l_{\mathcal{L}}(P) = \hat{h}_{\mathcal{L}}(nP) = n^2 \hat{h}_{\mathcal{L}}(P) = \tfrac{1}{2} n^2 b_{\mathcal{L}}(P,P) + n^2 l_{\mathcal{L}}(P),$$

for any integer n. Hence $l_{\mathcal{L}} = 0$.

If $\mathcal{L}$ is ample then $[-1]^*\mathcal{L}$ is ample too. Hence, $\mathcal{M} = \mathcal{L} \otimes [-1]^*\mathcal{L}$ is ample. Moreover by uniqueness

$$b_{\mathcal{L}} = \tfrac{1}{2} b_{\mathcal{L}} + \tfrac{1}{2} b_{[-1]^*\mathcal{L}} = \tfrac{1}{2} b_{\mathcal{M}}.$$

Therefore it suffices to prove that $b_{\mathcal{L}}$ is positive definite on $X(\overline{K}) \otimes \mathbb{R}$ for any symmetric ample invertible sheaf $\mathcal{L}$.

Since then $\hat{h}_{\mathcal{L}} = \frac{1}{2} b_{\mathcal{L}}$, it follows from Theorem 2.4 that for any finitely generated subgroup A of $X(\overline{K})$ and for any $C \in \mathbb{R}$ the cardinality of the set

$$\{P \in A \mid b_{\mathcal{L}}(P,P) \le C\}$$

is finite. It is not difficult to prove that this implies that $b_{\mathcal{L}}$ is positive definite on $X(\overline{K}) \otimes \mathbb{R}$. □

3.7 Lemma. *Let G be an abelian group and $h: G \to \mathbb{R}$ a function such that*

$$h(P+Q+R) - h(P+Q) - h(P+R) - h(Q+R) + h(P) + h(Q) + h(R) \sim 0,$$

as functions on G^3. Then there exists a unique symmetric bilinear mapping $b: G \times G \to \mathbb{R}$ and a unique homomorphism $l: G \to \mathbb{R}$ such that $h \sim \hat{h}$, where

$$\hat{h}(P) = \tfrac{1}{2} b(P,P) + l(P).$$

4 Metrized Line Bundles

The main interest of this section is to show that metrized line bundles on projective R-schemes give, in a natural way, a height function (i.e., not just a class of functions modulo bounded functions, but a specific element of such a class). For example, the Néron-Tate height can be obtained really nicely in this way, see for example [50], Chapter III.

For the moment let X be a complex analytic variety (i.e., a complex analytic space as in [27], App. B, which is reduced and Hausdorff), $\mathcal{E}$ a coherent $\mathcal{O}_X$-module (e.g., a locally free $\mathcal{O}_X$-module of finite rank). We denote by $\mathcal{C}_X^0$ the sheaf of continuous complex valued functions on X. If $U \subset X$ is an open subset and $f \in \mathcal{C}_X^0(U)$ then $\overline{f}$ denotes the complex conjugate of f (i.e., $\iota \circ f$, where $\iota: \mathbb{C} \to \mathbb{C}$ is the complex conjugation). We define a (continuous) hermitean form on $\mathcal{E}$ to be a morphism of sheaves $\langle \cdot, \cdot \rangle: \mathcal{E} \oplus \mathcal{E} \to \mathcal{C}_X^0$, such that:

(i) $\langle \cdot, \cdot \rangle$ is bi-additive,
(ii) $\langle f s_1, s_2 \rangle = f \langle s_1, s_2 \rangle$ for all $U \subset X$, $f \in \mathcal{O}_X(U)$, $s_1, s_2 \in \mathcal{E}(U)$,
(iii) $\langle s_2, s_1 \rangle = \overline{\langle s_1, s_2 \rangle}$ for all $U \subset X$, $s_1, s_2 \in \mathcal{E}(U)$.

A hermitean form is called a hermitean metric if moreover it is positive definite: $\langle s, s \rangle(x) > 0$ for all $U \subset X$, $s \in \mathcal{E}(U)$ and $x \in U$ such that $s(x) \neq 0$. The norm associated to a hermitean metric on $\mathcal{E}$ is the morphism of sheaves $\|\cdot\|: \mathcal{E} \to \mathcal{C}_X^0$, $s \mapsto \langle s, s \rangle^{1/2}$. Of course a hermitean metric is determined by its norm (use some polarization identity).

If $\mathcal{L}$ is an invertible $\mathcal{O}_X$-module then the norms of hermitean metrics on it can be easily characterized: they are the maps of sheaves $\|\cdot\|: \mathcal{L} \to \mathcal{C}_X^0$ such that $\|fs\| = |f|\,\|s\|$ and $\|s\|(x) > 0$ if $s(x) \neq 0$. In this case we will also call the norm associated to a hermitean metric a hermitean metric.

One can define differentiable, C^∞ or real analytic hermitean forms on $\mathcal{E}$ just by replacing the sheaf $\mathcal{C}_X^0$ by the sheaf of that kind of functions.

Let $\mathcal{E}' := \mathcal{C}_X^0 \otimes_{\mathcal{O}_X} \mathcal{E}$; then $\mathcal{E}'$ is a coherent $\mathcal{C}_X^0$-module. For any $\mathcal{C}_X^0$-module $\mathcal{E}'$ we define $\overline{\mathcal{E}'} := \mathcal{C}_X^0 \otimes_{\iota, \mathcal{C}_X^0} \mathcal{E}'$ and for any open $U \subset X$, $s \in \mathcal{E}'(U)$ we let $\overline{s} := 1 \otimes s \in \overline{\mathcal{E}'}(U)$. The maps $s \mapsto \overline{s}$ give an anti-linear isomorphism $\mathcal{E}' \to \overline{\mathcal{E}'}$ (i.e., one has $\overline{fs} = 1 \otimes fs = \overline{f} \otimes s = \overline{f} \cdot \overline{s}$). With these definitions, giving a hermitean form $\langle \cdot, \cdot \rangle$ on a coherent $\mathcal{O}_X$-module $\mathcal{E}$ is the same as giving a morphism of $\mathcal{C}_X^0$-modules $\phi: \mathcal{E}' \otimes_{\mathcal{C}_X^0} \overline{\mathcal{E}'} \to \mathcal{C}_X^0$, such that $\phi(s_1 \otimes \overline{s_2}) = \overline{\phi(s_2 \otimes \overline{s_1})}$. Using this description it is immediate that for $Y \to X$ a morphism of complex analytic varieties one has the notion of pullback for hermitean forms. In particular, for all $x \in X$ we get a hermitean form $\langle \cdot, \cdot \rangle(x)$ on the $\mathbb{C}$-vector space $\mathcal{E}(x)$ (the fibre of $\mathcal{E}$ at x, not the stalk). These $\langle \cdot, \cdot \rangle(x)$ vary continuously in the sense that if one expresses everything in coordinates then the coefficients of the matrix obtained that way are continuous. Of course, to give a continuous $\langle \cdot, \cdot \rangle$ is the same as to give a continuous family of $\langle \cdot, \cdot \rangle(x)$'s.

4.1 Example. The formula $\langle f, g \rangle = f\overline{g}$ defines a hermitean metric on $\mathcal{O}_X$. □

Suppose now that K is a number field and that $f: X \to \operatorname{Spec}(K)$ is a variety over K. Then for every $\sigma: K \to \mathbb{C}$ we have the variety X_σ over $\mathbb{C}$ obtained from $f: X \to \operatorname{Spec}(K)$ by pullback via $\operatorname{Spec}(\sigma): \operatorname{Spec}(\mathbb{C}) \to \operatorname{Spec}(K)$. Taking $\mathbb{C}$-valued points then gives for all σ a complex analytic variety $X_\sigma(\mathbb{C}) = \{P: \operatorname{Spec}(\mathbb{C}) \to X \mid f \circ P = \operatorname{Spec}(\sigma)\}$. If we denote $\overline{\sigma} = \iota \circ \sigma$ then we have maps $P \mapsto \overline{P} := P \circ \operatorname{Spec}(\iota)$:

$X_\sigma(\mathbb{C}) \to X_{\overline{\sigma}}(\mathbb{C})$; these maps are anti-holomorphic isomorphisms. For $\mathcal{E}$ a coherent $\mathcal{O}_X$-module let $\mathcal{E}_\sigma$ denote the coherent $\mathcal{O}_{X_\sigma(\mathbb{C})}$-module induced on $X_\sigma(\mathbb{C})$; for $U \subset X$ and $s \in \mathcal{E}(U)$ let $s_\sigma \in \mathcal{E}_\sigma(U_\sigma(\mathbb{C}))$ denote the section induced by s.

4.2 Definition. A hermitean metric on $\mathcal{E}$ is a set of hermitean metrics $\langle\cdot,\cdot\rangle_\sigma$ on the $\mathcal{E}_\sigma$, $\sigma: K \to \mathbb{C}$, such that for all open $U \subset X$, $s, t \in \mathcal{E}(U)$, $\sigma: K \to \mathbb{C}$ and $P \in U_\sigma(\mathbb{C})$ one has

$$\langle s_{\overline{\sigma}}, t_{\overline{\sigma}}\rangle_{\overline{\sigma}}(\overline{P}) = \overline{\langle s_\sigma, t_\sigma\rangle_\sigma(P)}$$

□

Another way to phrase the last condition is as follows: if we denote by $\mathcal{E}'$ the $\mathcal{C}^0_{X(\mathbb{C})}$-module on $X(\mathbb{C}) := \{P: \mathrm{Spec}(\mathbb{C}) \to X\} = \coprod_\sigma X_\sigma(\mathbb{C})$ induced by $\mathcal{E}$, and by $F_\infty: X(\mathbb{C}) \to X(\mathbb{C})$ the map $P \mapsto \overline{P}$, then one has $F_\infty^*\langle s,t\rangle = \overline{\langle s,t\rangle}$ for all $U \subset X$, $s, t \in \mathcal{E}(U)$. For our purposes we do not need this condition, but we impose it anyway in order to be compatible with the existing literature (e.g., [76]).

With our conventions concerning hermitean metrics on invertible sheaves we get the following definition.

4.3 Definition. A hermitean metric on an invertible $\mathcal{O}_X$-module $\mathcal{L}$ is a set of hermitean metrics $\|\cdot\|_\sigma: \mathcal{L}_\sigma \to \mathcal{C}^0_{X_\sigma(\mathbb{C})}$, $\sigma: K \to \mathbb{C}$, such that $\|s_{\overline{\sigma}}\|_{\overline{\sigma}}(\overline{P}) = \|s_\sigma\|_\sigma(P)$ for all $U \subset X$, $s \in \mathcal{L}(U)$, σ and $P \in U_\sigma(\mathbb{C})$. □

4.4 Example. Let $X := \mathbb{P}^n_K$, $\mathcal{L} := \mathcal{O}_X(m)$ and denote by $x_0, x_1, \ldots, x_n$ the homogeneous coordinates on X. By definition, a local section s of $\mathcal{L}$ can be written as F/G, where F and G in $K[x_0, x_1, \ldots, x_n]$ are homogeneous and $\deg(F) - \deg(G) = m$. The following formula defines a hermitean metric on $\mathcal{L}$:

$$\left\|\frac{\sigma(F)}{\sigma(G)}\right\|_\sigma (a_0 : a_1 : \cdots : a_n) = \frac{|(\sigma F)(a_0, a_1, \ldots, a_n)|}{|(\sigma G)(a_0, a_1, \ldots, a_n)|} \cdot \frac{1}{(|a_0|^2 + |a_1|^2 + \cdots + |a_n|^2)^{m/2}}$$

where $\sigma: K \to \mathbb{C}$ and $(a_0, a_1, \ldots, a_n) \in \mathbb{C}^{n+1}$. This metric will be our standard metric on $\mathcal{O}_{\mathbb{P}^n}(m)$. It is characterized, up to scalar multiple, by the property that at every σ it is invariant under the natural action of the unitary group $\mathrm{U}(n+1)$. □

As before, let R be the ring of integers in K. Let X be an integral projective R-scheme and let $\mathcal{L}$ be an invertible $\mathcal{O}_X$-module. By a hermitean metric on $\mathcal{L}$ we will mean a hermitean metric on the pullback $\mathcal{L}_K$ of $\mathcal{L}$ via $X_K := X \times_{\mathrm{Spec}(R)} \mathrm{Spec}(K) \to X$. By a metrized line bundle on X we will mean an invertible $\mathcal{O}_X$-module $\mathcal{L}$ with a given hermitean metric $\|\cdot\|$.

In particular, these definitions apply to Spec(R) itself. Let $(\mathcal{L}, \|\cdot\|)$ be a metrized line bundle on Spec(R), let $M := \Gamma(\mathrm{Spec}(R), \mathcal{L})$ and let $s \in M$ be non-zero. Then the degree of $(\mathcal{L}, \|\cdot\|)$ is the real number:

(4.4.1) $$\deg(\mathcal{L}) = \log \#(M/Rs) - \sum_{\sigma: K \to \mathbb{C}} \log \|s_\sigma\|_\sigma$$

(one easily checks that the right hand side is independent of s).

Let us now explain how a metrized line bundle $(\mathcal{L}, \|\cdot\|)$ on an integral projective R-scheme X gives a height function on the set $X(\overline{K})$ of $\overline{K}$-rational points of the variety X_K over K. Let $P \in X(\overline{K})$. Then there is a finite extension K' of K such that P is defined over K', i.e., P comes from a unique point $P \in X(K')$. Let R' be the ring of integers of K'. Then P can be uniquely extended to an R'-valued point (still denoted P) $P\colon \mathrm{Spec}(R') \to X$ by the valuative criterion of properness (see [27], Thm. 4.7). By pullback via P we get a hermitean line bundle $P^*\mathcal{L}$ on Spec(R'). The absolute height of P with respect to $(\mathcal{L}, \|\cdot\|)$ is then:

$$h_{(\mathcal{L},\|\cdot\|)}(P) := [K' : \mathbb{Q}]^{-1} \deg(P^*\mathcal{L}) \tag{4.4.2}$$

(one easily verifies that this is independent of K'). The following theorem states that $h_{(\mathcal{L},\|\cdot\|)}$ is in the class of height functions associated to $\mathcal{L}$ by Thm. 2.3.

4.5 Theorem. *If $(\mathcal{L}, \|\cdot\|)$ is a metrized line bundle on X then $h_{(\mathcal{L},\|\cdot\|)}$ and $h_{\mathcal{L}}$ are equivalent as functions on $X(\overline{K})$.*

Proof. It suffices to prove the theorem for $\mathcal{L}$ very ample, so we assume that X is a closed subscheme of $\mathbb{P}^n_R$ and that $\mathcal{L}$ is the restriction of $\mathcal{O}(1)$ to X. In virtue of Lemma 4.6 below we may choose a convenient hermitean metric on $\mathcal{L}$. Let us then choose the metric given by the formula:

$$\left\|\frac{\sigma(F)}{\sigma(G)}\right\|_\sigma (a_0 : a_1 : \cdots : a_n) = \frac{|(\sigma F)(a_0, a_1, \ldots, a_n)|}{|(\sigma G)(a_0, a_1, \ldots, a_n)|} \cdot \frac{1}{\max_{i\leq n} |a_i|}$$

(compare with Example 4.4). We claim that for this metric $h_{(\mathcal{L},\|\cdot\|)}$ is just the standard absolute height on $\mathbb{P}^n(\overline{K})$ of 1.1.1. In order to verify this for $P \in X(K')$, where K' is a finite extension of K, we may base change to the ring of integers R' of K'. Hence it is sufficient to check our claim for all K and all $P \in X(K)$.

We will compute deg $P^*\mathcal{O}(1)$, where P is an R-rational point of $X \subseteq \mathbb{P}^n_R$. We may assume that P^*x_0 is a nonzero section of $\Gamma(\mathrm{Spec}\, R, P^*\mathcal{O}(1))$. Then,

$$P^*\mathcal{O}(1)/RP^*x_0 \cong (\textstyle\sum Rx_i(P))/Rx_0(P) \cong (\sum R\frac{x_i}{x_0}(P))/R.$$

Hence,

$$\begin{aligned}\#P^*\mathcal{O}(1)/RP^*x_0 &= \prod_{v\notin M_K^\infty} \max_{i\leq n} \|\frac{x_i}{x_0}(P)\|_v \\ &= \left(\prod_{v\notin M_K^\infty} \max_{i\leq n} \|x_i(P)\|_v\right) \cdot \prod_{v\in M_K^\infty} \|x_0(P)\|_v.\end{aligned}$$

Therefore,

$$\begin{aligned}\deg P^*\mathcal{O}(1) &= \sum_{v\notin M_K^\infty} \log \max_{i\leq n} \|x_i(P)\|_v + \sum_{v\in M_K^\infty} \log \|x_0(P)\|_v \\ &\quad - \sum_{v\in M_K^\infty} \log \frac{\|x_0(P)\|_v}{\max_{i\leq n} \|x_i(P)\|_v} \\ &= [K\colon \mathbb{Q}] h_{\mathcal{L}}(P_K)\end{aligned}$$

□

4.6 Lemma. *Let $\mathcal{L}$ be a line bundle on X. If $\|\cdot\|$ and $\|\cdot\|'$ are hermitean metrics on $\mathcal{L}$ then there exist real numbers $c_1, c_2 > 0$ such that*

$$c_1\|s\|(P) \leq \|s\|'(P) \leq c_2\|s\|(P)$$

for any local section s of $\mathcal{L}$ and any $P \in X(\mathbb{C})$ where s is defined.

Proof. Since X is a projective R-scheme, $X(\mathbb{C})$ is a compact analytic variety. Therefore the continuous functions $\|\cdot\|'/\|\cdot\|$ and $\|\cdot\|/\|\cdot\|'$ on it are bounded. □

Chapter VI

D. Mumford's "A Remark on Mordell's Conjecture"

by Jaap Top

Throughout this exposition K will denote a number field and C/K will be an absolutely irreducible (smooth and complete) curve of genus $g \geq 1$. In fact more generally one can take for K any *field equipped with a product formula* as defined in [70, p. 7]. We make the standing assumption that over K, a divisor class of degree 1 on C exists (this can always be achieved after replacing K by a finite extension). Our main reference is the paper [53] referred to in the title above plus the descriptions given in [70], [9] of Mumford's result.

1 Definitions

Fix once and for all a divisor class a on C of degree 1. The jacobian $J = J(C)$ of C is an abelian variety of dimension g defined over K which represents divisor classes of degree 0 on C (see [14, p. 168] for a formal definition). In particular, the class a gives rise to an injective morphism

$$j = j_a \; : \; C \hookrightarrow J \; : \; P \mapsto (P) - a$$

defined over K. A. Weil proved in 1928 (case K a number field) that $J(K)$ is a finitely generated abelian group. This raises the possibility to study the set $C(K)$ of K-rational points on C as a subset of $J(K)$. The first successful attempt along this line seems to be due to Chabauty (1941). He combined the idea of considering $C(K)$ as a subset of a finitely generated abelian group with Skolem-type p-adic analysis and proved:

If $g = \text{genus}(C) > \text{rank}(J(K))$ then $C(K)$ is finite.

(cf. [70, §5.1] for a sketch of the proof; Coleman's paper (1985) [13] for effective bounds on the number of solutions in some cases.)

All other attempts starting from this basic setup seem to involve heights. Depending on a one defines a divisor

$$\Theta = \underbrace{j(C) + \ldots + j(C)}_{g-1} \subset J$$

with divisor class $\theta \in \mathrm{Pic}(J)$. To it one associates a canonical height h_θ on $J(\overline{K})$ and a bilinear form

$$\langle x, y\rangle = h_\theta(x+y) - h_\theta(x) - h_\theta(y).$$

Since θ is ample (this follows from the fact that θ defines a principal polarization on J; cf. [14, p. 186, Thm. 6.6 and p. 117, Prop. 9.1]), $\langle\cdot,\cdot\rangle$ is positive definite on $J(\overline{K})$/torsion. To verify this, note that $\langle x, x\rangle = \frac{1}{2}h_{\theta+[-1]^*\theta}(x)$; now θ is ample, hence $\theta + [-1]^*\theta$ is ample and symmetric. The rest of the argument is a trivial exercise (see [70, §3.7]).

The existence of this non-degenerate bilinear form was discovered by Néron and by Tate (mid-60's). Combined with Weil's result it gives $J(K)\otimes\mathbb{R}$ the structure of a euclidean space E. An easy consequence is

$$\#\{x \in J(K);\ ||x|| \le T\} = (\text{pos. const.})\cdot T^{\mathrm{rk}} + o(T^{\mathrm{rk}}).$$

2 Mumford's Inequality

The previous section raises the possibility to study rational points on C by means of the sequence

$$C(K) \longrightarrow J(K)/\text{torsion} \subset E$$

in which the first map is finite to one (in fact, by Raynaud's proof of the Manin-Mumford conjecture [60] there is a bound on the number of points in the fibres independent of the field K); and the second map gives $J(K)$/torsion as a lattice in euclidean space E. Mumford's result on this which we want to discuss here can be seen as a first step towards Vojta's proof (simplified by Bombieri) of Mordell's conjecture (1989/90) and even to Faltings's theorem which is the topic of this book. The result is

2.1 Theorem (Mumford, 1965). *With the notations as above, there is a real number $c > 0$ such that for all $x \neq y \in C(\overline{K})$ the inequality*

$$||x||^2 + ||y||^2 - 2g\langle x, y\rangle \ge -c(1 + ||x|| + ||y||)$$

holds.

3 Interpretation, Consequences and an Example

Since for $g < 2$ the left hand side is non-negative, Mumford's inequality is at most interesting for $g \ge 2$. In that case, for K a number field, if one considers points in $C(K)$ only then the result is trivially implied by the truth of Mordell's conjecture. Needless to say, this will not be used here.

One way to interpret the inequality is as follows. Suppose $0 < r < \sqrt{g^2-1}$. Then for $x \in C(\overline{K})$ with $||x|| >> 0$, no $y \in C(\overline{K})$ is in $B(gx, r||x||)$:= the ball with center gx and radius $r||x||$. Indeed, were $y \in C(\overline{K})$ inside this ball, then of course $||y|| \le (r+g)||x||$ and

$$\begin{aligned} 0 &\ge ||y - gx||^2 - r^2||x||^2 \ge -c - c||x|| - c||y|| + (g^2 - 1 - r^2)||x||^2 \\ &\ge -c - c(1 + r + g)||x|| + (g^2 - 1 - r^2)||x||^2 \end{aligned}$$

which implies that $||x||$ is small.

Mumford's inequality implies

3.1 Corollary. *There is a real number c_2 such that*

$$\#\{x \in C(K);\ ||x|| \leq T\} \leq c_2 \log T.$$

The same result (with of course a different c_2) is true with $||\cdot||$ replaced by any $h = h_\delta$ on C; with δ a divisor of positive degree. Comparing the above result with the "density" of rational points in $J(K)$ one says that the set $C(K) \subset J(K)$ is "widely spaced" (this notion is formally introduced by Silverman [75]; also Serre uses it [70, p. 105], in fact already in the French version from 1980 (p. 2.7: "très espacés")).

Proof. (of Cor 3.1). Take $c_3 = 5c + 1/2$, then for $x, y \in C(K)$ with $2||x|| \geq ||y|| \geq ||x|| \geq c_3$ one has

$$||x||^2 + ||y||^2 \geq c_3||x|| + c_3||y|| \geq 5c(1 + ||x|| + ||y||),$$

hence

$$\frac{\langle x, y\rangle}{||x|| \cdot ||y||} \leq \frac{||x||^2 + ||y||^2 + c(1 + ||x|| + ||y||)}{2g||x|| \cdot ||y||} \leq \frac{3}{5g}\left(\frac{||x||}{||y||} + \frac{||y||}{||x||}\right) \leq \frac{3}{2g}.$$

Now take $T >> 0$. The $x \in C(K)$ with $||x|| \leq T$ can be subdivided in the x's with $||x|| \leq c_3$ and the ones with $c_3 2^j < ||x|| \leq c_3 2^{j+1}$ (for $0 \leq j \leq c_4 \log T$). In one such "interval" different points satisfy $\langle ||x||^{-1}x, ||y||^{-1}y\rangle \leq 3/(2g)$. The maximal number of unit vectors in E satisfying this angle constraint is easily seen to be bounded in terms of $\dim E = \operatorname{rank} J(K)$, say bounded by c_5. Then

$$\#\{x \in C(K);\ ||x|| \leq T\} \leq \#\text{torsion in } J(K) \cdot (\#\{||x|| \leq c_3\} + c_5 c_4 \log T) \leq c_2 \log T.$$

□

3.2 Example. Take C, D curves defined over $\mathbb{F}_q$ such that a non-constant morphism $D \to C$ over $\mathbb{F}_q$ exists. Put $K = \mathbb{F}_q(D)$. A point $x \in C(K)$ corresponds to a morphism $\varphi_x : D \to C$ defined over $\mathbb{F}_q$ and for $h(x)$ we can take $h(x) = \deg(\varphi_x)$. Let π denote the Frobenius homomorphism on K. Clearly with x also $\pi^n x \in C(K)$, and $h(\pi^n x) = q^n h(x)$. Hence in this example

$$\#\{x \in C(K);\ h(x) \leq T\} \geq (\text{pos. const.}) \cdot \log T.$$

□

4 The Proof assuming some Properties of Divisor Classes

Let D be a (smooth, complete, irreducible) curve of genus g over an algebraically closed field k. Write J_D for its jacobian and use as above a divisor class a of degree 1 to define j and θ. Denote by p_i the projection $J_D \times J_D \to J_D$ onto the ith factor and let $s : J_D \times J_D \to J_D$ be summation ($s(x, y) = x + y$). Furthermore define $a' \in J_D$ by $a' = (2g-2)a - K_D$ in which K_D is the canonical class on D. Lastly, $\tau_b : J_D \to J_D$ will denote translation over $b \in J_D$. The following properties of divisor classes are used in the proof of Mumford's result:

1. $\theta' := [-1]^*\theta = \tau^*_{-a'}\theta$.

2. $j^*\theta' = ga$.

3. $(j \times j)^*(s^*\theta - p_1^*\theta - p_2^*\theta) = a \times D + D \times a - \Delta$.

Theorem 2.1 is obtained from this as follows. We return to the situation in §§1–2 and we will use the statements above in this context. Write $h_\theta(x) = \frac{1}{2}\langle x, x\rangle + \ell(x)$; here ℓ is linear. Then $\frac{1}{2}\langle x,x\rangle - \ell(x) = h_\theta(-x) = h_{\theta'}(x) = h_{\tau^*_{-a'}\theta}(x) = h_\theta(x - a') = \frac{1}{2}\langle x,x\rangle - \langle x, a'\rangle + \frac{1}{2}\langle a', a'\rangle + \ell(x - a') = \frac{1}{2}\langle x,x\rangle + \ell(x) - \langle x, a'\rangle$ (here use that $h_\theta(0) = 0$). The conclusion is that

$$\ell(x) = \frac{1}{2}\langle x, a'\rangle.$$

Next, using heights on $C \times C$, identifying x with $j(x)$ (as we already did a couple of times) and ignoring the (bounded) functions arising from the fact that we choose actual heights depending on a divisor class:

$$\begin{aligned} h_\Delta(x,y) &= h_{a\times C}(x,y) + h_{C\times a}(x,y) - h_{(j\times j)^*s^*\theta}(x,y) + h_{(j\times j)^*p_1^*\theta}(x,y) \\ &\quad + h_{(j\times j)^*p_2^*\theta}(x,y) \\ &= h_a(x) + h_a(y) - h_\theta(x+y) + h_\theta(x) + h_\theta(y) = \frac{1}{g}h_{\theta'}(x) + \frac{1}{g}h_{\theta'}(y) - \langle x, y\rangle \\ &= \frac{1}{2g}\left\{||x||^2 + ||y||^2 - 2g\langle x,y\rangle - \langle x, a'\rangle - \langle y, a'\rangle\right\}. \end{aligned}$$

Since Δ is effective, h_Δ is bounded from below on $\{x \neq y\}$, hence the result follows.

5 Proof of the Divisorial Properties

Proof. (of (1)) The class θ is by definition represented by the divisor

$$\Theta = \left\{[\sum_{i=1}^{g-1} P_i - (g-1)a];\ P_i \in D\right\}$$

hence θ' by $\Theta' = \{[(g-1)a - \sum P_i]\}$. Therefore it suffices to show that for general $\sum_{i=1}^{g-1} P_i$ a unique $\sum_{i=1}^{g-1} Q_i$ exists such that $\sum P_i + \sum Q_i$ is canonical. This follows from Riemann-Roch: one has $\ell(\sum P_i) - \ell(K_D - \sum P_i) = 0$. Since $\ell(\sum P_i) \neq 0$ (constant morphisms are in L(any effective divisor)), it follows that also the class of $K_D - \sum P_i$ is represented by an effective divisor. A slightly more geometric argument runs as follows. In case D is not hyperelliptic one may assume that D is canonically embedded; then the canonical divisors are precisely the hyperplane sections, and $g - 1$ points determine a hyperplane and therefore a canonical divisor. For hyperelliptic D, write ι for the hyperelliptic involution. Then $\sum Q_i = \sum \iota P_i$ is canonical. □

Proof. (of (2)) Take $c \in \mathrm{Pic}^0(D)$ generic and write $b = a - c$. The embedding of D into J_D using b instead of a will be denoted j_b. We will show $j_b^*\theta' = ga - c$ which by specialization implies the desired formula.

For $x \in D$ to satisfy $j_b(x) \in \Theta'$ is equivalent to $x + \sum_{i=1}^{g-1} P_i \sim ga - c$ (with $\sim$ denoting linear equivalence). Now $ga - c$ is a generic divisor of degree g, hence by

Riemann-Roch its divisor class can be written as $\sum_{i=1}^{g} Q_i$ in a unique way (in fact, the gth symmetric power of D is birational to J_D). In other words, the possible choices for $\{x, \{\sum P_i\}\}$ are just the partitions of $\{Q_i\}$ in two sets with 1 and $g-1$ elements, respectively. Hence the divisor of these x's is $\sum Q_i \sim ga - c$. □

Proof. **(of (3))** Recall the seesaw principle [14, §5, p. 109-110] (which we only state in the special case $D \times D$):

> *If $\delta \in Pic(D \times D)$ is trivial restricted to all vertical and one horizontal fibre, then $\delta = 0$.*

Apply this to $\delta = (j \times j)^*(s^*\theta - p_1^*\theta - p_2^*\theta) + \Delta - a \times D - D \times a$. This is a symmetric divisor class, hence it suffices to show $\delta|_{x \times D}$ is trivial. Consider $\varphi = \varphi_x \;:\; D \to D \times D \;:\; y \mapsto (x, y)$. Clearly $\varphi^*(\Delta - a \times D - D \times a) \sim x - 0 - a = x - a$. Furthermore, $p_1 \circ (j \times j) \circ \varphi = j$, $p_2 \circ (j \times j) \circ \varphi = \text{constant}$ and $s \circ (j \times j) \circ \varphi = \tau_{x-a} \circ j$. Hence in Pic($D$):

$$\varphi^*\delta = j^*\tau_{x-a}^*\theta - j^*\theta + x - a.$$

Now for any $b \in \mathrm{Pic}^0(D)$ one has

$$\begin{aligned} j^*\tau_b^*\theta &= j^*\tau_b^*\tau_{a'}^*\theta' \text{ (use (1))} = j_{-b+K_D-(2g-1)a}^*\theta' = \\ &= ga - (2g-2)a - b + K_D \text{ (from the proof of (2))} = K_D - b - (g-2)a. \end{aligned}$$

It follows that $\varphi^*\delta = (K_D - x + a - (g-2)a) - (K_D - (g-2)a) + x - a = 0$ as required. □

6 Effectiveness and Generalizations

In what follows, K is a number field. One of Paul Vojta's ideas was to replace the quadratic part $||x||^2 + ||y||^2 - 2g\langle x, y\rangle$ in Mumford's result by more generally $\lambda||x||^2 + \mu||y||^2 - 2g\nu\langle x, y\rangle$, with λ, μ, ν suitably chosen, depending on x, y. This leads to considering an other divisor in $C \times C$, namely

$$\lambda(a \times C) + \mu(C \times a) - \nu(j \times j)^*(s^*\theta - p_1^*\theta - p_2^*\theta) \sim (\lambda - \nu)a \times C + (\mu - \nu)C \times a + \nu\Delta.$$

To obtain a lower bound for this, we need firstly that this divisor class is effective. Secondly we want that a positive representative of this class does not contain (x, y). In fact, this second condition can be weakened to "does not contain (x, y) with very high multiplicity". Since one wants to choose λ, μ, ν and hence the divisor class depending on (x, y), this is a problem. Vojta, Faltings and Bombieri each showed us a different way to cope with this difficulty, and the final result is

6.1 Fact. *There exists $c' > 1$ such that for $x, y \in C(\overline{K})$ with $||x|| \geq c'$ and $||y||/||x|| \geq c'$ one has*

$$\langle x, y\rangle \leq \frac{3}{4}||x|| \cdot ||y||.$$

Note that this easily implies Mordell's conjecture.

It should be remarked that both c and c' can be chosen independent of the field K. They only depend on the curve C, on the divisor a and on chosen embeddings of C and of $C \times C$ into projective space, using fixed very ample divisors Na and $M_1(a \times C) + M_2(C \times a)$, respectively. This information yields a radius ρ, an angle $\alpha > 0$ and a number n, such that for any number field $L \supseteq K$ the points in $C(L)$ can be described as follows: there are the ones within the ball $B(0, \rho)$, plus each cone with angle α contains outside this ball at most n other points.

Chapter VII

Ample Line Bundles and Intersection Theory

by Johan de Jong

1 Introduction

This chapter gives an overview of the results from intersection theory we need for the proof of Faltings's theorem. Meanwhile we will try to give a coherent account of (this part of) intersection theory and we will try to show what a beautiful theory it is. We would like to stress here, that it should be possible for anyone with some basic knowledge of (algebraic) geometry to prove all the results mentioned in this chapter after reading the first 40 pages of Hartshorne's Lecture Notes [28].

2 Coherent Sheaves, etc.

X will always denote a projective variety over the algebraically closed field k. An *affine open subvariety of* X is an open $U \subset X$ such that U is isomorphic as a variety to a closed subvariety $Z \subset \mathbb{A}_k^N$ for some N.

2.1 Example. If X is a closed subvariety of some $\mathbb{P}^n$ with homogeneous coordinates $T_0, \ldots, T_n$ then for any non-zero homogeneous polynomial $P(T_0, \ldots, T_n)$ the set

$$\{(t_0 : \cdots : t_n) \in X(k) \mid P(t_0, \ldots, t_n) \neq 0\}$$

is an affine open subvariety of X. □

The affine open subsets form a basis for the topology on X and hence a sheaf on X is determined by the sets of sections over the affine open $U \subset X$. The *structure sheaf* $\mathcal{O}_X$ of X is determined by the rule

$$\mathcal{O}_X(U) = \Gamma(U, \mathcal{O}_X) = \Gamma(U, \mathcal{O}_U) = k[T_1, \ldots, T_N]/I$$

if $U \subset X$ is isomorphic to the affine variety determined by the (prime) ideal $I \subset k[T_1, \ldots, T_N]$.

The other sheaves occurring in this chapter will always be *sheaves of* $\mathcal{O}_X$*-modules*, that is, they will be sheaves of abelian groups $\mathcal{F}$ on X endowed with a multiplication

$\mathcal{O}_X \times \mathcal{F} \to \mathcal{F}$ such that for each open $U \subset X$, $\mathcal{F}(U)$ becomes a module over $\mathcal{O}_X(U)$. A sheaf of $\mathcal{O}_X$-modules $\mathcal{F}$ is said to be *generated by global sections* if there exists a family of sections $(s_i)_{i\in I}$, $s_i \in \Gamma(X, \mathcal{F})$ such that the map of sheaves

$$\oplus_{i\in I}\mathcal{O}_X \longrightarrow \mathcal{F}, \qquad \oplus_{i\in I} f_i \mapsto \sum_i f_i s_i$$

is surjective (i.e., any local section of $\mathcal{F}$ should be locally a finite linear combination of the s_i).

A *coherent* sheaf of $\mathcal{O}_X$-modules is a sheaf $\mathcal{F}$ such that

- for each open affine $U \subset X$, $\mathcal{F}(U)$ is a finitely generated $\mathcal{O}_X(U)$-module,
- if $U \subset V \subset X$ are open affine then $\mathcal{F}(U) = \mathcal{F}(V) \otimes_{\mathcal{O}_X(V)} \mathcal{O}_X(U)$

The fundamental theorem on sheaves/coherent sheaves is

2.2 Theorem. *For any sheaf $\mathcal{F}$ on X, $H^i(X, \mathcal{F}) = 0$ for all $i > \dim X$. If $\mathcal{F}$ is coherent, then $H^i(X, \mathcal{F})$ is a finite dimensional k-vector space for all i.*

This theorem allows us to define the *Euler-Poincaré characteristic* of $\mathcal{F}$:

$$\chi(\mathcal{F}) = \sum_{i=0}^{\infty}(-1)^i \dim_k H^i(X, \mathcal{F}).$$

Using the long exact cohomology sequence, we see that for an exact sequence of coherent sheaves $0 \to \mathcal{F}_1 \to \mathcal{F}_2 \to \mathcal{F}_3 \to 0$ we have $\chi(\mathcal{F}_2) = \chi(\mathcal{F}_1) + \chi(\mathcal{F}_3)$.

2.3 Examples.

(a) Ideal sheaves. An ideal sheaf is a subsheaf $\mathcal{I} \subset \mathcal{O}$ such that for each $U \subset X$ open, $\mathcal{I}(U)$ is an ideal of $\mathcal{O}_X(U)$. It is coherent, since for each open affine $U \subset X$ the ring $\mathcal{O}_X(U)$ is a Noetherian ring.

An ideal sheaf $\mathcal{I}$ determines a closed subset Z of X

$$Z = \{x \in X \mid \text{for all } x \in U \subset X \text{ and } f \in \mathcal{I}(U)\text{: } f(x) = 0\}$$

and it determines a sheaf of rings $\mathcal{O}_Z = \mathcal{O}_X/\mathcal{I}$. The pair $(Z, \mathcal{O}_Z)$ is (what we will call) a *closed subscheme* of X. It is usually denoted by Z.

A *closed subvariety* is a closed subscheme $(Z, \mathcal{O}_Z)$ such that the sheaf of algebras $\mathcal{O}_Z$ is without zero divisors.

Example: $X = \mathbb{A}^1_k$, $\Gamma(X, \mathcal{O}_X) = k[t] \supset I = (t^n)$. In this case we get $(Z_n, \mathcal{O}_{Z_n})$ with $Z_n = \{0\}$ and $\Gamma(X, \mathcal{O}_{Z_n}) = k[t]/(t^n)$. Hence $Z_n \neq Z_{n'}$ unless $n = n'$, and only Z_1 is a closed subvariety.

If $\mathcal{F}$ is a coherent sheaf on X then we can construct an ideal sheaf $\mathcal{I}$ as follows:

$$\mathcal{I}(U) = \{f \in \mathcal{O}_X(U) \mid \text{multiplication by } f\colon \mathcal{F}|_U \to \mathcal{F}|_U \text{ is the zero map}\}$$

The closed subscheme Z associated to this is called the *support* of $\mathcal{F}$: $Z := \mathrm{supp}(\mathcal{F})$.

Example: The support of $\mathcal{O}_Z$ is $(Z, \mathcal{O}_Z)$.

(b) Line bundles. A *line bundle* or *invertible sheaf* is a coherent sheaf $\mathcal{L}$ on X such that each point $x \in X$ has an affine neighborhood $U \subset X$ such that $\mathcal{L}(U) \cong \mathcal{O}_X(U)$ as $\mathcal{O}_X(U)$-modules, i.e., $\mathcal{L}|_U \cong \mathcal{O}_X|_U$.

If we have two line bundles $\mathcal{L}_1$ and $\mathcal{L}_2$ then we can form the tensor product $\mathcal{L}_1 \otimes_{\mathcal{O}_X} \mathcal{L}_2$. It is again a line bundle. Using the tensor product for multiplication, the set of line bundles up to isomorphism becomes an abelian group with identity $\mathcal{O}_X$ and inverse $\mathcal{L}^{-1} = \mathbf{Hom}_{\mathcal{O}_X}(\mathcal{L}, \mathcal{O}_X)$. This group is denoted by $\mathrm{Pic}(X)$, it is the *Picard group* of X. For morphisms $f: X \to Y$ there is a *pullback* $f^*: \mathrm{Pic}(Y) \to \mathrm{Pic}(X)$ defined as follows: if the line bundle $\mathcal{L}$ on Y is trivial on the open sets U_i which cover Y and if its transition functions are $f_{i,j} \in \Gamma(U_i \cap U_j, \mathcal{O}_Y^*)$ then $f^*\mathcal{L}$ is the line bundle on X which is trivial on the open sets $f^{-1}U_i$ and has transition functions $f_{i,j} \circ f \in \Gamma(f^{-1}(U_i \cap U_j), \mathcal{O}_X^*)$. The pullback is a homomorphism of abelian groups. □

3 Ample and Very Ample Line Bundles

Recall that a line bundle $\mathcal{L}$ on X is said to be *very ample* if it is generated by global sections and for a basis $s_0, \ldots, s_n$ of $\Gamma(X, \mathcal{L})$ the map $\phi_{s_0,\ldots,s_n}: X \to \mathbb{P}_k^n$ is a closed immersion. The line bundle $\mathcal{L}$ is said to be *ample* if there exist $N \in \mathbb{N}$ such that $\mathcal{L}^{\otimes N} := \mathcal{L} \otimes \cdots \otimes \mathcal{L}$ (N factors $\mathcal{L}$) is very ample. A cohomological criterion for being ample is the following.

3.1 Theorem (Serre-Grothendieck). *For a line bundle $\mathcal{L}$ on X the following are equivalent:*

1. *$\mathcal{L}$ is ample,*
2. *for every coherent sheaf $\mathcal{F}$ on X we have $H^i(X, \mathcal{F} \otimes \mathcal{L}^{\otimes n}) = 0$ for all $i > 0$, $n \gg 0$,*
3. *for every coherent sheaf $\mathcal{F}$ on X, the sheaf $\mathcal{F} \otimes \mathcal{L}^{\otimes n}$ is generated by global sections for $n \gg 0$.*

3.2 Corollary. *Let $f: X \to Y$ be a finite morphism of projective varieties and $\mathcal{L}$ a line bundle on Y. If $\mathcal{L}$ is ample then $f^*\mathcal{L}$ is ample on X. If f is surjective and $f^*\mathcal{L}$ is ample then $\mathcal{L}$ is ample.*

The proof uses the theorem above and comparison of cohomology of sheaves on X and on Y.

3.3 Corollary. *For any line bundle $\mathcal{L}$ on X there exist very ample line bundles $\mathcal{L}_1$ and $\mathcal{L}_2$ such that $\mathcal{L} \cong \mathcal{L}_1 \otimes \mathcal{L}_2^{-1}$.*

Proof. Suppose $\mathcal{M}$ is very ample on X. For some $n \in \mathbb{N}$ the line bundle $\mathcal{L} \otimes \mathcal{M}^{\otimes n}$ is generated by global sections. It is easy to see that $\mathcal{L}_1 = \mathcal{L} \otimes \mathcal{M}^{\otimes n} \otimes \mathcal{M}$ and $\mathcal{L}_2 = \mathcal{M}^{\otimes(n+1)}$ are very ample (and $\mathcal{L} \cong \mathcal{L}_1 \otimes \mathcal{L}_2^{-1}$). □

4 Intersection Numbers

Let $\mathcal{F}$ be a coherent sheaf on X and let $\mathcal{L}_1, \ldots, \mathcal{L}_t$ be line bundles on X.

4.1 Proposition. *The function $(n_1, \ldots, n_t) \mapsto \chi(\mathcal{L}_1^{\otimes n_1} \otimes \cdots \otimes \mathcal{L}_t^{\otimes n_t} \otimes \mathcal{F})$ is a numerical polynomial of total degree at most $\dim(\mathrm{supp}(\mathcal{F}))$.*

Proof. We prove the proposition by induction on $\dim(\operatorname{supp}\mathcal{F})$. If $\dim(\operatorname{supp}\mathcal{F}) = 0$ then we have $\mathcal{L}_1^{\otimes n_1} \otimes \cdots \otimes \mathcal{L}_t^{\otimes n_t} \otimes \mathcal{F} \cong \mathcal{F}$ since each $\mathcal{L}_i$ is trivial in a neighborhood of the support of $\mathcal{F}$. Further, $H^i(X, \mathcal{F}) = 0$ unless $i = 0$ since $\mathcal{F}$ is supported in dimension zero. Hence our function is constant and equal to $\chi(\mathcal{F}) = \dim_k H^0(X, \mathcal{F})$. For the induction step, by Cor. 3.3 above, we may assume that each $\mathcal{L}_i$ is very ample. Hence we can choose a section $s \in \Gamma(X, \mathcal{L}_i)$ whose zero set is "transversal" to $\operatorname{supp}(\mathcal{F})$, i.e., such that $\dim((s = 0) \cap \operatorname{supp}(\mathcal{F})) < \dim(\operatorname{supp}\mathcal{F})$. Thus we see that the kernel and cokernel of multiplication by s:

$$0 \to \mathcal{K} \to \mathcal{F} \xrightarrow{s} \mathcal{F} \otimes \mathcal{L}_i \to \mathcal{Q} \to 0$$

are sheaves with lower dimensional support. Hence for all i we have that

$$\chi(\mathcal{L}_1^{\otimes n_1} \otimes \cdots \otimes \mathcal{L}_i^{\otimes (n_i+1)} \otimes \cdots \otimes \mathcal{L}_t^{\otimes n_t} \otimes \mathcal{F}) - \chi(\mathcal{L}_1^{\otimes n_1} \otimes \cdots \otimes \mathcal{L}_t^{\otimes n_t} \otimes \mathcal{F})$$

is a polynomial in $n_1, \ldots, n_t$ of total degree at most $\dim(\operatorname{supp}\mathcal{F}) - 1$. From this the result follows. □

4.2 Definition. If $(Z, \mathcal{O}_Z)$ is a closed subscheme of dimension d and $\mathcal{L}_1, \ldots, \mathcal{L}_d$ are line bundles then we define $(\mathcal{L}_1 \cdot \cdots \cdot \mathcal{L}_d \cdot Z)$, the *intersection number* of Z with $\mathcal{L}_1, \ldots, \mathcal{L}_d$, to be the coefficient of $n_1 \cdots n_d$ in the polynomial $\chi(\mathcal{L}_1^{\otimes n_1} \otimes \cdots \otimes \mathcal{L}_d^{\otimes n_d} \otimes \mathcal{O}_Z)$. □

4.3 Remarks.

(a) The intersection number is an integer (any numerical polynomial f of degree $\leq d$ in $n_1, \ldots, n_d$ can be written uniquely as a $\mathbb{Z}$-linear combination of the functions $\binom{n_1}{k_1} \cdot \cdots \cdot \binom{n_d}{k_d}$ with $k_i \geq 0$ for all i and $\sum_i k_i \leq d$).

(b) It is additive:

$$(\mathcal{L}_1 \otimes \mathcal{L}_1' \cdot \cdots \cdot \mathcal{L}_d \cdot Z) = (\mathcal{L}_1 \cdot \cdots \cdot \mathcal{L}_d \cdot Z) + (\mathcal{L}_1' \cdot \cdots \cdot \mathcal{L}_d \cdot Z)$$

and symmetric: the intersection number is independent of the order of $\mathcal{L}_1, \ldots, \mathcal{L}_d$.

(c) Suppose that the irreducible components Z_i of Z with $\dim Z_i = \dim Z$ have generic points $z_1, \ldots, z_k$. Then the rings $\mathcal{O}_{Z,z_i}$ (= localisation of $\mathcal{O}_Z$ at z_i) are zero dimensional and of finite length. The multiplicity of Z_i in Z is defined as $m_{Z_i,Z}$ = length of $\mathcal{O}_{Z,z_i}$. One can show that

$$(\mathcal{L}_1 \cdot \cdots \cdot \mathcal{L}_d \cdot Z) = \sum_{i=1}^{k} m_{Z_i,Z} \cdot (\mathcal{L}_1 \cdot \cdots \cdot \mathcal{L}_d \cdot Z_i)$$

This one proves by showing that $\mathcal{O}_Z$ has a filtration by coherent sheaves

$$\mathcal{O}_Z = \mathcal{F}_0 \supset \mathcal{F}_1 \supset \cdots \supset \mathcal{F}_l$$

with $l = \sum_{i=1}^k m_{Z_i,Z}$, such that for all j with $0 \leq j < l$ there exists i in $\{1, \ldots, k\}$ and morphisms

$$\mathcal{F}_j/\mathcal{F}_{j+1} \longrightarrow \mathcal{G}_j \longleftarrow \mathcal{H}_j \longrightarrow \mathcal{O}_{Z_i}$$

with kernel and cokernel supported in dimension smaller than d.

(d) The proof of the proposition above also gives a method for computing $(\mathcal{L}_1 \cdot \cdots \cdot \mathcal{L}_d \cdot Z)$. Suppose the section $s \in \Gamma(X, \mathcal{L}_d)$ is such that

$$\cdot s \colon \mathcal{L}_d^{-1} \otimes \mathcal{O}_Z \to \mathcal{O}_Z \qquad (*)$$

is *injective*. In this case the quotient sheaf $\mathcal{O}_Z/s\mathcal{O}_Z$ is the structure sheaf of a closed subscheme which we will denote by $Z \cap H_d$, where $H_d = \{x \in X \mid s(x) = 0\}$, and we get

$$(\mathcal{L}_1 \cdot \cdots \cdot \mathcal{L}_d \cdot Z) = (\mathcal{L}_1 \cdot \cdots \cdot \mathcal{L}_{d-1} \cdot (H_d \cap Z))$$

Remarks. (1) H_d is a divisor representing $\mathcal{L}_d$, i.e., such that $\mathcal{L}_d = \mathcal{O}(H_d)$
(2) If $\mathcal{L}$ is very ample, then $(*)$ is injective for general s.
Repeating this, we get that $(\mathcal{L}_1 \cdot \cdots \cdot \mathcal{L}_d \cdot Z)$ is equal to the number of points in the intersection $H_1 \cap \cdots \cap H_d \cap Z$ counted with multiplicities, if the divisors $H_1, \ldots, H_d$ are chosen general enough:

$$(\mathcal{L}_1 \cdot \cdots \cdot \mathcal{L}_d \cdot Z) = \sum_{P \in H_1 \cap \cdots \cap H_d \cap Z} e_P(H_1, \ldots, H_d; Z)$$

By the proof of the proposition above, the local multiplicity of $H_1 \cap \cdots \cap H_d \cap Z$ at P is:

$$e_P(H_1, \ldots, H_d; Z) = \dim_k \mathcal{O}_{X,P} / (\mathcal{I}_{Z,P} + f_1\mathcal{O}_{X,P} + \cdots + f_d\mathcal{O}_{X,P})$$

here $f_1, \cdots, f_d$ are local equations defining $H_1, \ldots, H_d$.
(e) The number $(\mathcal{L}^d \cdot Z) := (\mathcal{L} \cdot \cdots \cdot \mathcal{L} \cdot Z)$ (d factors $\mathcal{L}$) can also be computed by the rule:

$$\chi(\mathcal{L}^{\otimes n} \otimes \mathcal{O}_Z) = (\mathcal{L}^d \cdot Z)(d!)^{-1} \cdot n^d + \text{lower order terms}$$

If $\mathcal{L}$ is ample we will call $(\mathcal{L}^d \cdot Z)$ the *degree of Z with respect to $\mathcal{L}$* and write $\deg_{\mathcal{L}} Z$ for it.

Suppose now that $X = \mathbb{P}^n$ and $Z \subset \mathbb{P}^n$ is an irreducible subvariety of dimension $d > 0$. By remark (d):

$$\deg_{\mathcal{O}(1)} Z = \#\,((\text{general } n{-}d\text{-plane}) \cap Z) = \deg Z,$$

the classical definition of the degree of Z. Using the fact that any hyperplane in $\mathbb{P}^n$ intersects Z non-trivially and induction on d we see that we always have that $\deg Z > 0$. This more or less proves one direction of the following theorem.

4.3.1 Theorem (Nakai criterion). *A line bundle $\mathcal{L}$ on X is ample if and only if for every closed subvariety $Z \subset X$ we have $(\mathcal{L}^d \cdot Z) > 0$.*

(f) The behavior of the degree under a finite morphism $f \colon X \to Y$ is the following:

$$\deg_{f^*\mathcal{L}} Z = \deg(f \colon Z \to f(Z)) \cdot \deg_{\mathcal{L}} f(Z)$$

Here $Z \subset X$ is a closed subvariety and $\mathcal{L}$ is an ample line bundle on Y.
Sketch of proof and explanation of $\deg(f \colon Z \to f(Z))$. The image $f(Z)$ of Z under f is again a closed subvariety. The map f defines an inclusion $f^* \colon k(f(Z)) \hookrightarrow k(Z)$ of function fields and this is a finite field extension. Put $n := \deg(f \colon Z \to f(Z)) := [k(Z) : k(f(Z))]$. If this field extension is separable, then generically for $p \in$

$f(Z)$ we have $f^{-1}(p) = \{q_1, \ldots, q_n\} \subset Z$. Hence, if $H_1 \cap \cdots \cap H_d \cap f(Z) = \{p_1, \ldots, p_m\}$ then $f^*H_1 \cap \cdots \cap f^*H_d \cap Z = \{q_{11}, \ldots, q_{1n}, \ldots, q_{m1}, \ldots, q_{mn}\}$. The result follows. □

(g) Suppose that $k = \mathbb{C}$ and that X is smooth. Then $X(\mathbb{C})$ also has the structure of a complex manifold, which we will denote by X^{an}. A line bundle $\mathcal{L}$ on X gives rise to a line bundle $\mathcal{L}^{an}$ on X^{an} which is determined by an element of $H^1(X^{an}, \mathcal{O}^*_{an})$. The exponential sequence:

$$0 \to 2\pi i\mathbb{Z} \to \mathcal{O}_{an} \stackrel{\exp}{\to} \mathcal{O}^*_{an} \to 0$$

gives us the first Chern class $c_1(\mathcal{L}) \in H^2(X^{an}, \mathbb{Z})$. On the other hand, an irreducible subvariety $Z \subset X$ will give rise to an analytic subvariety $Z^{an} \subset X^{an}$. After triangulizing it, we see that it gives rise to a class $[Z] = [Z^{an}] \in H_{2d}(X^{an}, \mathbb{Z})$. For a general closed subscheme Z we put

$$[Z] := \sum_i m_{Z_i,Z} \cdot [Z_i^{an}] \in H_{2d}(X^{an}, \mathbb{Z})$$

The connection between intersection numbers and cup product in cohomology is now:

$$(\mathcal{L}_1 \cdot \cdots \cdot \mathcal{L}_d \cdot Z) = \langle c_1(\mathcal{L}_1) \wedge \cdots \wedge c_1(\mathcal{L}_d), [Z] \rangle$$

where $\langle\ ,\ \rangle : H^{2d}(X^{an}, \mathbb{Z}) \times H_{2d}(X^{an}, \mathbb{Z}) \to \mathbb{Z}$ is the canonical pairing. □

5 Numerical Equivalence and Ample Line Bundles

We write $F_1(X) = \{\sum n_i C_i \mid C_i \subset X \text{ is a closed subvariety of dimension } 1\}$ for the free abelian group generated by the set of all curves in X. By abuse of language we call an element of $F_1(X)$ a *curve* on X. A curve $\sum n_i C_i$ is *effective* if $n_i \geq 0$ for all i. Our intersection number gives us a bilinear pairing

$$\text{Pic}(X) \times F_1(X) \to \mathbb{Z} \qquad (\mathcal{L}, \sum n_i C_i) \mapsto \sum n_i (\mathcal{L} \cdot C_i)$$

We say that two curves C_1, C_2 (resp. line bundles $\mathcal{L}_1$, $\mathcal{L}_2$) are *numerically equivalent* if $(\mathcal{L} \cdot C_1) = (\mathcal{L} \cdot C_2)$ for all $\mathcal{L} \in \text{Pic}(X)$ (resp. $(\mathcal{L}_1 \cdot C) = (\mathcal{L}_2 \cdot C)$ for all $C \in F_1(X)$). **Notation:** $C_1 \equiv C_2$ (resp. $\mathcal{L}_1 \equiv \mathcal{L}_2$).

5.1 Definition. $A^1(X) := (\text{Pic}(X)/\equiv) \otimes_{\mathbb{Z}} \mathbb{R}$ and $A_1(X) := (F_1(X)/\equiv) \otimes_{\mathbb{Z}} \mathbb{R}$. (By definition the intersection number gives a non-degenerate pairing between these two $\mathbb{R}$-vector spaces.) □

5.2 Remark. If $k = \mathbb{C}$ and X is smooth then by Remark 4.3(g) we know that the intersection products $(\mathcal{L} \cdot C)$ only depend on the first Chern class $c_1(\mathcal{L}) \in H^2(X^{an}, \mathbb{Z})$. Hence we conclude that $A^1(X)$ is a subquotient of $H^2(X^{an}, \mathbb{R})$ (it is actually a subspace!). This proves the following general theorem in this special case. □

5.3 Theorem. *$A^1(X)$ is finite dimensional.*

In general one reduces to the case where X is a smooth surface and then the theorem is a consequence of the Mordell-Weil theorem for abelian varieties over function fields.

5.4 Remark. If $\mathcal{L}_1 \equiv \mathcal{L}_1'$ then $(\mathcal{L}_1 \cdot \mathcal{L}_2 \cdot \dots \cdot \mathcal{L}_d \cdot Z) = (\mathcal{L}_1' \cdot \mathcal{L}_2 \cdot \dots \cdot \mathcal{L}_d \cdot Z)$ always. Hence, by the Nakai criterion, whether or not $\mathcal{L}$ is ample depends only on the numerical equivalence class of $\mathcal{L}$; i.e., it depends only on the point in $A^1(X)$ determined by $\mathcal{L}$. □

A subset S of a $\mathbb{R}$-vector space V is called a cone if for all x, y in S and for all real numbers $a > 0$ and $b > 0$, one has $ax + by \in S$. Let $A_1^+(X) \subset A_1(X)$ be the cone generated by the classes of effective curves. We know that if $\mathcal{L}$ is ample then $(\mathcal{L} \cdot C) > 0$ for all $C \in A_1^+(X)$. Hence the cone P^o (the *ample cone*) spanned by the classes of ample line bundles is contained in the *pseudo-ample cone*

$$P = \{D \in A^1(X) \mid D \cdot C \geq 0 \ \forall C \in A_1^+(X)\}$$

We will have to use the following basic theorem once:

5.5 Theorem. *P^o is the interior of P. (The ample cone is the interior of the pseudo-ample cone.)*

6 Lemmas to be used in the Proof of Thm. I of Faltings

Suppose X is an abelian variety over k and $\mathcal{L}$ is an ample line bundle on X. We denote by $[n]$ the morphism given by multiplication by n on X: $[n]: X \to X$, $x \mapsto nx$.

6.1 Lemma. $[n]^*\mathcal{L} \equiv \mathcal{L}^{\otimes n^2}$.

Proof. (for $k = \mathbb{C}$) We can write $X^{an} = \mathbb{C}^g/L$ and $L \cong \mathbb{Z}^{2g}$. The induced map $[n]^*: H^2(X^{an}, \mathbb{Z}) \to H^2(X^{an}, \mathbb{Z})$ is equal to multiplication by n^2, as is seen from the identification $H^2(X^{an}, \mathbb{Z}) \cong \bigwedge^2 H^1(X^{an}, \mathbb{Z}) \cong \bigwedge^2 L^*$. □

Suppose we have a product situation $X = X_1 \times X_2$. Let us denote by $p_1: X \to X_1$ the projection onto X_1. Further, we assume given line bundles $\mathcal{L}_1, \dots, \mathcal{L}_k$ on X_1 and line bundles $\mathcal{L}_{k+1}, \dots, \mathcal{L}_d$ on X. Finally, $Z \subset X$ is a closed subvariety of dimension d.

6.2 Lemma. *If $\dim p_1(Z) < k$ then $(p_1^*(\mathcal{L}_1) \cdot \dots \cdot p_1^*(\mathcal{L}_k) \cdot \mathcal{L}_{k+1} \cdot \dots \cdot \mathcal{L}_d \cdot Z) = 0$.*

Proof. By linearity of the intersection product and of p_1^* we may assume $\mathcal{L}_1, \dots, \mathcal{L}_k$ are very ample. By our assumption we can find divisors $H_1, \dots, H_k$ (divisors of sections of $\mathcal{L}_1, \dots, \mathcal{L}_k$) such that $H_1 \cap \dots \cap H_k \cap p_1(Z) = \emptyset$ (recall that $\dim p_1(Z) < k$). Hence also $p_1^*H_1 \cap \dots \cap p_1^*H_k \cap Z = \emptyset$. This implies by Remark 4.3(d) that

$$(p_1^*(\mathcal{L}_1) \cdot \dots \cdot \mathcal{L}_d \cdot Z) = (\mathcal{L}_{k+1} \cdot \dots \cdot \mathcal{L}_d \cdot (p_1^*H_1 \cap \dots \cap p_1^*H_k \cap Z)) = (\mathcal{L}_{k+1} \cdot \dots \cdot \mathcal{L}_d \cdot \emptyset) = 0$$

□

Put $P := \mathbb{P}^{n_1} \times \dots \times \mathbb{P}^{n_m}$. On it we have the line bundles $\mathcal{L}_i := \mathrm{pr}_i^*(\mathcal{O}_{\mathbb{P}^{n_i}}(1))$.

6.3 Exercises.

(a) Sections $F \in \Gamma(P, \mathcal{L}_1^{\otimes d_1} \otimes \cdots \otimes \mathcal{L}_m^{\otimes d_m})$ correspond with multihomogeneous polynomials of multidegree $(d_1, \ldots, d_m)$.

(b) $\mathrm{Pic}(P) = \mathbb{Z}[\mathcal{L}_1] \oplus \cdots \oplus \mathbb{Z}[\mathcal{L}_m]$ and $A^1(P) = \mathbb{R}^m$ (same basis). The pseudo-ample cone is $\{(x_1, \ldots, x_m) \in \mathbb{R}^m \mid \forall i: x_i \geq 0\}$, the ample cone is $\{(x_1, \ldots, x_m) \in \mathbb{R}^m \mid \forall i: x_i > 0\}$. If $Z \subset P$ is an irreducible subvariety, then we get degrees indexed by $e_1, \ldots, e_m$ with $\sum e_i = \dim(Z)$: the numbers $(\mathcal{L}_1^{e_1} \cdot \mathcal{L}_2^{e_2} \cdot \cdots \cdot \mathcal{L}_m^{e_m} \cdot Z)$. These numbers are ≥ 0; use induction on $\dim Z$ and the fact that one can always find a section in $\mathcal{L}_i$ "transversal" to Z. □

6.4 Lemma. (Prop. 2.3 of [22].) *Suppose $X \subset P$ is a closed subscheme which is an intersection of hypersurfaces of multidegree $(d_1, \ldots, d_m)$. If the X_j are irreducible components of X with multiplicities m_j, and of the same codimension t, then*

$$\sum m_j(\mathcal{L}_1^{e_1} \cdot \cdots \cdot \mathcal{L}_m^{e_m} \cdot X_j) \leq \left(\mathcal{L}_1^{e_1} \cdot \cdots \cdot \mathcal{L}_m^{e_m} \cdot (\mathcal{L}_1^{\otimes d_1} \otimes \cdots \otimes \mathcal{L}_m^{\otimes d_m})^t \cdot P\right)$$

Proof. The difficult point in this lemma is the fact that the X_j do not need to be the irreducible components of X of the maximal dimension. The proof is by induction on t ($t = 0$ is trivial). First we choose t polynomials $F_1, \ldots, F_t$ of multidegree $(d_1, \ldots, d_m)$ in the ideal of X, such that each X_j is an irreducible component of their set of common zeros. (Having chosen $F_1, \ldots, F_k$ one simply chooses for F_{k+1} a polynomial which is non-zero on all the components of $V(F_1) \cap \cdots \cap V(F_k)$ which contain an X_j.) Then we might as well replace X by the closed subscheme defined by $F_1, \ldots, F_t$ (this will only make the m_j bigger) and enlarge the set of X_j to include all the components of X of codimension t.

Let us denote by Y the closed subscheme defined by $F_1, \ldots, F_{t-1}$ and by Y_i the components of Y of codimension $t-1$ containing some X_j. The multiplicity of Y_i in Y is n_i. By our choice of F_t we have: the irreducible components of $(F_t = 0) \cap Y_i$ are the X_j say with multiplicities k_{ij}. The formula $m_j = \sum_i k_{ij} n_i$ is a consequence of the fact that the sequence $F_1, \ldots, F_t$ is a regular sequence in the local ring $\mathcal{O}_{P,x_j}$ ($x_j \in X_j$ a generic point). Hence we get:

$$\begin{aligned}
\sum_j m_j(\mathcal{L}_1^{e_1} \cdot \cdots \cdot \mathcal{L}_m^{e_m} \cdot X_j) &= \sum_{i,j} k_{ij} n_i (\mathcal{L}_1^{e_1} \cdot \cdots \cdot \mathcal{L}_m^{e_m} \cdot X_j) \\
&= \sum_i n_i (\mathcal{L}_1^{e_1} \cdot \cdots \cdot \mathcal{L}_m^{e_m} \cdot (F_t = 0) \cap Y_i) \\
&= \sum_i n_i (\mathcal{L}_1^{e_1} \cdot \cdots \cdot \mathcal{L}_m^{e_m} \cdot \mathcal{L} \cdot Y_i) \\
&\leq \text{(induction)} \leq (\mathcal{L}_1^{e_1} \cdot \cdots \cdot \mathcal{L}_m^{e_m} \cdot \mathcal{L} \cdot \mathcal{L}^{t-1} \cdot P)
\end{aligned}$$

here $\mathcal{L} = \mathcal{L}_1^{\otimes d_1} \otimes \cdots \otimes \mathcal{L}_m^{\otimes d_m}$. □

Chapter VIII

The Product Theorem

by Marius van der Put

1 Differential Operators and Index

In this chapter, k will be a field of characteristic zero.

1.1 Definition. A vector field on a variety X over k is a k-linear derivation D of the sheaf $\mathcal{O}_X$, i.e., $D: \mathcal{O}_X \to \mathcal{O}_X$ is k-linear and for all open subsets $U \subset X$ and all $f, g \in \mathcal{O}_X(U)$ one has:

$$D(fg) = D(f)g + fD(g)$$

In other words, a vector field on X is a global section of the sheaf $\mathbf{Hom}_{\mathcal{O}_X}(\Omega^1_{X/k}, \mathcal{O}_X)$ (see [27], II.8). □

1.2 Definition. Let $r \geq 0$. A differential operator on X of degree $\leq r$ is a k-linear endomorphism L of the sheaf $\mathcal{O}_X$ such that every $x \in X$ has an open neighborhood U on which L can be written as a sum of expressions $fD_1 \cdots D_s$ $(s \leq r)$, where each D_i is a vector field on U and $f \in \mathcal{O}_X(U)$. (The factor f can be omitted for $s > 0$.) □

1.3 Remark. This definition is not the same as the one in [25], IV, §16.8, but one can show that the two definitions are equivalent (note that k has characteristic zero). □

1.4 Example. Let $X = \mathbb{P}^n$ with homogeneous coordinates $x_0, \ldots, x_n$. Let E be any k-derivation of $k(x_0, \ldots, x_n)$ such that $E(x_i)$ is homogeneous linear for all i. Then E induces a vector field D on $\mathbb{P}^n$ where $D =$ the restriction of E to $k(x_1/x_0, \ldots, x_n/x_0)$. Indeed $D(x_i/x_j) = E(x_i/x_j) = (x_jE(x_i) - x_iE(x_j))/x_j^2 \in k[x_0/x_j, \ldots, x_n/x_j]$. A calculation shows that every vector field on $\mathbb{P}^n$ is obtained in this way. In affine coordinates $y_1 = x_1/x_0, \ldots, y_n = x_n/x_0$ the vector field D has the form:

$$\sum_{i=1}^{n} (\text{degree} \leq 1) \frac{\partial}{\partial y_i} + (\text{degree} \leq 1) \sum_{i=1}^{n} y_i \frac{\partial}{\partial y_i}$$

□

1.5 Example. Let $m \geq 1$. Define $\mathcal{P} = \mathbb{P}^{n_1} \times \cdots \times \mathbb{P}^{n_m}$ and let $x_{\bullet}(i)$ denote the homogeneous coordinates of $\mathbb{P}^{n_i}$. Let $R = k[x_{\bullet}(1), \ldots, x_{\bullet}(m)]$ be the (multi-)graded ring of $\mathcal{P}$. As in the previous example one can show that any vector field D on $\mathcal{P}$ is induced by a k-derivation E of R which respects the multigrading, i.e., $E(x_\bullet(i))$ is a k-linear combination of the elements $x_\bullet(i)$ and

$$D\left(\frac{x_\bullet(i)}{x_0(i)}\right) = \frac{x_0(i)E(x_\bullet(i)) - x_\bullet(i)E(x_0(i))}{x_0(i)^2}$$

A calculation shows that every differential operator L on $\mathcal{P}$ can globally be written as a polynomial in global vector fields; it follows that L acts on R and respects the multi-grading. □

Let $m \geq 1$, $X_1, \ldots, X_m$ varieties over k and $X := X_1 \times \cdots \times X_m$. Any vector field D on X can be written uniquely as $D = \sum_{i=1}^m D_i$, where D_i is a vector field on X in the direction of the i-th factor, i.e., the image of D_i under the projection $\mathrm{pr}_{j\neq i}\colon X \to \prod_{j\neq i} X_j$ is zero (note that $\Omega^1_{X/k} = \oplus_i \mathrm{pr}_i^* \Omega^1_{X_i/k}$).

1.6 Definition. A differential operator L on X is of multi-degree $\leq (r_1, \ldots, r_m)$ if, locally, it can be written as a sum of expressions $fD_1 \cdots D_s$, where the D_i are vector fields in the directions of the factors of X and for all j the number of D_i in the jth direction is at most r_j. □

Let $d_1, \ldots, d_m$ be positive integers and let $f \neq 0$ be a section of a line bundle $\mathcal{L}$ on X. In the following definition these d_i will be used to define a kind of weighted total degree for differential operators on X: the weighted degree of a vector field in the ith direction will be $1/d_i$.

1.7 Definition. A differential operator L on X is of weighted degree $\leq r$ (or $< r$) if, locally, it is a sum of differential operators of some multi-degree $\leq (r_1, \ldots, r_m)$ with $r_1/d_1 + \cdots + r_m/d_m \leq r$ (resp. $< r$). □

1.8 Definition. Let $x \in X$ and s a generator of the stalk $\mathcal{L}_x$ of $\mathcal{L}$ at x. Then $f = gs$ in $\mathcal{L}_x$ for a unique $g \in \mathcal{O}_{X,x}$. We define the index $i(x, f)$ of f at x, with respect to the weights $d_1, \ldots, d_m$, to be $\sigma \in \mathbb{Q}_{\geq 0}$ with σ maximal for the property:

> $(L(g))(x) = 0$ for every differential operator L defined on some neighborhood of x and of weighted degree $< \sigma$.

(This property of σ and g is easily seen to be independent of the choice of s.) □

Note that in this definition x is not necessarily a closed point of X. For x any point of X and g any element of $\mathcal{O}_{X,x}$ we define the value $g(x) \in k(x)$ of g at x by $g(x) := i_x^* g$, where $i_x\colon \mathrm{Spec}(k(x)) \to X$ is the inclusion.

1.9 Example. Let $X_i = \mathbf{A}^1$ for all i, $\mathcal{L} = \mathcal{O}_X$, $x = 0$ and

$$f = \sum_{(i_1,\ldots,i_m)} a_{i_1,\ldots,i_m} x_1^{i_1} \cdots x_m^{i_m}$$

Then

$$i(x, f) = \min\{i_1/d_1 + \cdots + i_m/d_m \mid a_{i_1,\ldots,i_m} \neq 0\}$$

□

1.10 Definition. In the situation of Def. 1.8, let $\sigma \in \mathbb{R}$. The closed subscheme Z_σ of X on which the index of f is at least σ is defined by the sheaf of ideals in $\mathcal{O}_X$ generated locally by the $L(g)$, where L is a differential operator of weighted degree $< \sigma$ and $f = gs$ with s a local generator of $\mathcal{L}$. □

1.11 Example. Let $\mathcal{P}$ be as above, let $e_1, \ldots, e_m$ be non-negative integers and let $\mathcal{L}$ be the line bundle $\mathcal{O}(e_1, \ldots, e_m)$. We have already seen that global differential operators on $\mathcal{P}$ act on $\Gamma(\mathcal{P}, \mathcal{O}(e_1, \ldots, e_m))$ and that the sheaf of differential operators of multi-degree $\leq (j_1, \ldots, j_m)$ is generated by its global sections. It follows that Z_σ is defined by the homogeneous ideal of R generated by the $L(f)$, where L is of multi-degree $\leq (j_1, \ldots, j_m)$ with $j_1/d_1 + \cdots + j_m/d_m < \sigma$. Note that all $L(f)$ are global sections of $\mathcal{O}(e_1, \ldots, e_m)$. □

2 The Product Theorem

2.1 Theorem. *Suppose that k is algebraically closed. Let $m, n_1, \ldots, n_m$ be positive integers. Let $\mathcal{P} := \mathbb{P}^{n_1} \times \cdots \times \mathbb{P}^{n_m}$. For every $\epsilon > 0$ there exists $r \in \mathbb{R}$ such that if:*

1. *$d_1, \ldots, d_m$ are positive integers satisfying $d_1/d_2 \geq r, \ldots, d_{m-1}/d_m \geq r$,*
2. *$f \in \Gamma(\mathcal{P}, \mathcal{O}(d_1, \ldots, d_m))$ is non-zero, and*
3. *for some σ, Z is an irreducible component of Z_σ and of $Z_{\sigma+\epsilon}$ (here the index of f is taken with respect to the weights $d_1, \ldots, d_m$)*

then:

(i) Z is a product of closed subvarieties Z_i of $\mathbb{P}^{n_i}$, i.e., $Z = Z_1 \times \cdots \times Z_m$,

(ii) the degrees $\deg(Z_i)$ are bounded in terms of ϵ and $n_1 + \cdots + n_m$ only.

Proof. Let $\mathrm{pr}_i : \mathcal{P} \to \mathbb{P}^{n_i}$ be the ith projection. Let $Z_i := \mathrm{pr}_i Z$ and $f_i := \dim Z_i$. Then Z_i is irreducible and closed in $\mathbb{P}^{n_i}$ and of course Z is contained in $Z_1 \times \cdots \times Z_m$. We have $Z = Z_1 \times \cdots \times Z_m$ if and only if $\sum f_i = \dim Z$. Let $\mathcal{L}_i$ be the line bundle $\mathrm{pr}_i^* \mathcal{O}(1)$ on $\mathcal{P}$ and consider the intersection numbers $\mathcal{L}_1^{e_1} \cdot \cdots \cdot \mathcal{L}_m^{e_m} \cdot Z$ for $m$tuples $(e_1, \ldots, e_m)$ of non-negative integers with $\sum e_i = \dim Z$ (see Chapter VII).

We claim that $\sum f_i = \dim Z$ if and only if there exists only one such $m$tuple $(e_1, \ldots, e_m)$ with $\mathcal{L}_1^{e_1} \cdot \cdots \cdot \mathcal{L}_m^{e_m} \cdot Z > 0$. Namely, if $\sum f_i = \dim Z$ and $e_i > f_i$ for some i, then $\mathcal{L}_1^{e_1} \cdot \cdots \cdot \mathcal{L}_m^{e_m} \cdot Z = 0$ because there exist e_i hyperplanes in $\mathbb{P}^{n_i}$ whose common intersection with Z_i is empty (see Chapter VII, Lemma 6.2). Suppose now that $\sum f_i > \dim Z$. There exist f_1 hyperplanes in $\mathbb{P}^{n_1}$ such that their common intersection with Z_1 is a non-empty set of dimension zero. It follows that there exist $e_1, \ldots, e_m$ with $e_1 = f_1$ and $\mathcal{L}_1^{e_1} \cdot \cdots \cdot \mathcal{L}_m^{e_m} \cdot Z \neq 0$. For some i we must have $e_i < f_i$. But in the same way we show the existence of $e'_1, \ldots, e'_m$ with $e'_i = f_i$ and $\mathcal{L}_1^{e'_1} \cdot \cdots \cdot \mathcal{L}_m^{e'_m} \cdot Z \neq 0$.

For what follows we need a lower bound for the multiplicity m_{Z,Z_σ} of Z in Z_σ. Recall (Chapter VII, §4) that m_{Z,Z_σ} is by definition the length of the local ring $\mathcal{O}_{Z_\sigma,\eta}$, where η is the generic point of Z. The lower bound for m_{Z,Z_σ} we want is a consequence of the assumption that Z is an irreducible component of both Z_σ and $Z_{\sigma+\epsilon}$.

First some notation. Let $\mathrm{pr}_{\geq i} : \mathcal{P} \to \prod_{j \geq i} \mathbb{P}^{n_j}$ and $\mathrm{pr}_{>i} : \mathcal{P} \to \prod_{j>i} \mathbb{P}^{n_j}$ denote the projections. Let $\delta_i := \dim(\mathrm{pr}_{\geq i} Z) - \dim(\mathrm{pr}_{>i} Z)$; hence $\delta_1 + \cdots + \delta_m = \dim Z$. Let $\mathcal{F}_i$

be the fiber of $(\mathrm{pr}_{\geq i}Z \to \mathrm{pr}_{>i}Z)$ over $\mathrm{pr}_{>i}\eta$, let $\eta_i := \mathrm{pr}_{\geq i}\eta$ be the generic point of $\mathcal{F}_i$ and let $k_i := k(\eta_i)$. Then $\mathcal{F}_i$ is a closed subvariety of $\mathbf{P}^{n_i}_{k_{i+1}}$ of dimension δ_i.

2.2 Lemma. *With the notations as above, suppose that $d_1 \geq \cdots \geq d_m$, then we have:*

$$m_{Z,Z_\sigma} \geq (\epsilon/\mathrm{codim}(Z))^{\mathrm{codim}(Z)} \prod_{i=1}^{m} d_i^{n_i-\delta_i}$$

Proof. We want to get $t_{i,j}$ ($1 \leq i \leq m$ and $1 \leq j \leq n_i$) in $\mathcal{O}_{\mathcal{P},\eta}$, such that:

(i) the $dt_{i,j}$ form a basis of the free $\mathcal{O}_{\mathcal{P},\eta}$-module $\Omega^1_{\mathcal{P}/k,\eta}$,

(ii) the $t_{i,j}$ with $j > \delta_i$ form a system of parameters of the regular local ring $\mathcal{O}_{\mathcal{P},\eta}$ of dimension codim(Z),

(iii) the images in $\Omega^1_{Z/k,\eta}$ of the $dt_{i,j}$ with $j \leq \delta_i$ form a basis of that free $\mathcal{O}_{Z,\eta}$-module,

(iv) each $t_{i,j}$ is a pullback via $\mathrm{pr}_{\geq i}$, i.e., $t_{i,j}$ is constant in the first $i-1$ directions.

Such a system $t_{i,j}$ can be gotten inductively, starting with $i = m$. Let $1 \leq i \leq m$ and suppose that we have already chosen the $t_{l,j}$ with $l > i$. Take $s_j \in \mathcal{O}_{\mathcal{F}_i,\eta_i}$, $1 \leq j \leq \delta_i$, such that the ds_j form a $\mathcal{O}_{\mathcal{F}_i,\eta_i}$-basis of $\Omega^1_{\mathcal{F}_i/k_{i+1},\eta_i}$. Let s_j, $\delta_i < j \leq n_i$, be a system of parameters of the regular local ring $\mathcal{O}_{\mathbf{P}^{n_i}_{k_{i+1}},\eta_i}$. Put $t_{i,j} := \mathrm{pr}_{\geq i}^{-1} s_j$.

Let the k-derivations $\partial_{i,j}\colon \mathcal{O}_{\mathcal{P},\eta} \to \mathcal{O}_{\mathcal{P},\eta}$ be the dual basis of the basis $dt_{i,j}$ of $\Omega^1_{\mathcal{P}/k,\eta}$. From the last property of the $t_{i,j}$ it follows that each $\partial_{i,j}$ is a linear combination of vector fields in the directions of the lth factors, with $l \leq i$ (the "$\leq$" is explained by the passage to the dual basis). It follows that $\partial_{i,j}$ is of weighted degree $\leq 1/d_i$, since we suppose that $d_1 \geq \cdots \geq d_m$. From now on we only consider $\partial_{i,j}$ with $j > \delta_i$. The number of those is codim(Z).

Let $I_\sigma \subset \mathcal{O}_{\mathcal{P}}$ and $I_{\sigma+\epsilon} \subset \mathcal{O}_{\mathcal{P}}$ be the ideal sheaves of Z_σ and $Z_{\sigma+\epsilon}$, respectively. By construction, we have $LI_\sigma \subset I_{\sigma+\epsilon} \subset I_Z$ for all differential operators L of weighted degree $\leq \epsilon$. In particular, this holds for all $L = \prod_{i,j} \partial_{i,j}^{\alpha_{i,j}}$ with $\alpha_{i,j} \leq \epsilon d_i/\mathrm{codim}(Z)$. Let α_i be the largest integer, less than or equal to $\epsilon d_i/\mathrm{codim}(Z)$. It follows that the ideal $I_{Z_\sigma,\eta}$ of $\mathcal{O}_{\mathcal{P},\eta}$ is contained in the ideal generated by the $t_{i,j}^{\alpha_i+1}$ ($1 \leq i \leq m$, $j > \delta_i$). Hence:

$$\mathrm{length}(\mathcal{O}_{Z_\sigma,\eta}) \geq \prod_{i=1}^{m} (\alpha_i+1)^{n_i-\delta_i} \geq (\epsilon/\mathrm{codim}(Z))^{\mathrm{codim}(Z)} \prod_{i=1}^{m} d_i^{n_i-\delta_i}$$

□

Let $(e_1, \ldots, e_m)$ be an $m$tuple of non-negative integers with $\sum_i e_i = \dim Z = \sum \delta_i$. According to Prop. 2.3 of [22] (see Lemma 6.4 of Chapter VII) and the lemma above, one has

$$\begin{aligned}(\mathcal{L}_1^{e_1} \cdot \cdots \cdot \mathcal{L}_m^{e_m} \cdot Z) &\leq \frac{1}{m_{Z,Z_\sigma}} \mathcal{L}_1^{e_1} \cdot \cdots \cdot \mathcal{L}_m^{e_m} \cdot (d_1\mathcal{L}_1 + \cdots + d_m\mathcal{L}_m)^{\mathrm{codim}(Z)} \\ &\leq (\mathrm{codim}(Z)/\epsilon)^{\mathrm{codim}(Z)} \prod_i d_i^{\delta_i-n_i} \frac{\mathrm{codim}(Z)!}{\prod_i (n_i-e_i)!} \prod_i d_i^{n_i-e_i} \\ &\leq c(\epsilon, \mathrm{codim}(Z)) \prod_{i=1}^{m} d_i^{\delta_i-e_i}\end{aligned}$$

with $c(\epsilon,t) = (t/\epsilon)^t t!$. For $1 \leq i \leq m$ let $\eta_i := \sum_{j=i}^{m}(\delta_j - e_j)$. Then $\eta_1 = 0$ and $\prod_{i=1}^{m} d_i^{\delta_i-e_i} = \prod_{i=2}^{m} (d_i/d_{i-1})^{\eta_i}$.

Suppose now that $\mathcal{L}_1^{e_1} \cdot \dots \cdot \mathcal{L}_m^{e_m} \cdot Z \neq 0$. Then for $1 \leq i \leq m$ one has:

$$e_i + \dots + e_m \leq \dim(\mathrm{pr}_{\geq i} Z) = \delta_i + \dots + \delta_m$$

It follows that $\eta_i \geq 0$ for all i. Finally suppose that

$$d_i / d_{i+1} \geq r > \max(c(\epsilon, \mathrm{codim}(Z)), 1)$$

Let $\eta := \eta_2 + \dots + \eta_m$. Then we have:

$$1 \leq \mathcal{L}_1^{e_1} \cdot \dots \cdot \mathcal{L}_m^{e_m} \cdot Z \leq c(\epsilon, \mathrm{codim}(Z)) r^{-\eta}$$

So we find that $\eta = 0$. It follows that $e_i = \delta_i$ for all i, so by the argument at the beginning of the proof one has $Z = Z_1 \times \dots \times Z_m$ and $\dim(Z_i) = \delta_i$ for all i.

To prove the second part of the theorem, note that

$$\deg(Z_1) \cdot \dots \cdot \deg(Z_m) = \mathcal{L}_1^{\delta_1} \cdot \dots \cdot \mathcal{L}_m^{\delta_m} \cdot Z \leq c(\epsilon, \mathrm{codim}(Z))$$

It follows that $\deg(Z_i) \leq c(\epsilon, \mathrm{codim}(Z))$ for all i. □

2.3 Remark. The Product theorem will be used in the following way. Let N be an integer greater than $\dim(\mathcal{P})$. Suppose that f has index at least σ at a point x. Then there exists a chain $\mathcal{P} \neq Z_1 \supset Z_2 \supset \dots \supset Z_N \ni x$, with Z_i an irreducible component of $Z_{i\sigma/N}$. It follows that for some i one has $Z_i = Z_{i+1}$. Then one applies the Product theorem with $\epsilon = \sigma/N$. □

2.4 Remark. Suppose that (with the terminology of the Product theorem) f is defined over some subfield k_0 of k. Then Z_σ and $Z_{\sigma+\epsilon}$ are also defined over k_0. Let d be the number of conjugates of Z under the Galois group of k over k_0. Applying Prop. 2.3 of [22] (see Lemma 6.4 of Chapter VII) to these conjugates of Z one finds $d \leq c(\epsilon, \mathrm{codim}(Z))$. It follows that Z is defined over an extension k_1 of k_0 with $[k_1 : k_0] \leq c(\epsilon, \mathrm{codim}(Z))$. □

3 From the Product Theorem to Roth's Lemma

This section will not be used in the rest of the book, but is included to show, following Vojta in [81], §18, how one derives Roth's lemma from the following arithmetic version of the Product theorem (Thm. 3.3 of [22]).

Let f, ϵ, r, $d_1, \dots, d_m$ and Z be as in the Product Theorem. Suppose that f and Z are defined over $\mathbb{Q}$ and choose an affine product $\mathbf{A}^{n_1} \times \dots \times \mathbf{A}^{n_m} \subset \mathcal{P}$ such that $Z \cap \mathbf{A}^{n_1} \times \dots \times \mathbf{A}^{n_m} \neq \emptyset$. On $\mathbf{A}^{n_1} \times \dots \times \mathbf{A}^{n_m}$ we represent f by a non-zero $F \in \mathbb{Z}[y_\cdot(1), \dots, y_\cdot(m)]$. Let $\log|F| := \log(\max|\text{coefficients of } F|)$ denote the logarithmic height of F. Then one has:

1) $Z = Z_1 \times \dots \times Z_m$

2) $\deg(Z_i) \leq c(\epsilon, \mathrm{codim}(Z))$

3) $\sum_{i=1}^{m} d_i h(Z_i) \leq c_1(\epsilon) \log|F| + c_2(\epsilon)(d_1 + \dots + d_m)$

In this statement $h(T)$ denotes the logarithmic height of a subscheme T of $\mathbb{P}^N_{\mathbb{Z}}$. We will give neither definitions nor proofs. We note that for $p, q \in \mathbb{Z}$ with $\gcd(p,q) = 1$ the subscheme "$px_0 + qx_1 = 0$" of $\mathbb{P}^1_{\mathbb{Z}}$ has logarithmic height $\log(\max|p|, |q|)$.

3.1 Lemma (Roth). *Let $m \geq 1$ and $\epsilon > 0$. There exist positive numbers r, C_2 and C_1 such that if:*
(i) *$d_1, \ldots, d_m$ satisfy $d_i/d_{i+1} \geq r$ for all i,*
(ii) *$q_1, \ldots, q_m$ are positive integers such that $\forall i$: $q_i^{d_i} \geq q_1^{d_1}$, $\log q_i \geq C_2$,*
(iii) *$p_1, \ldots, p_m$ are integers with $\gcd(p_i, q_i) = 1$,*
(iv) *$F \in \mathbb{Z}[y_1, \ldots, y_m]$, $F \neq 0$, satisfies $|F|^{C_1} \leq q_1^{d_1}$ and has, for all i, degree at most d_i in y_i,*
then $\text{index}_{(d_1,\ldots,d_m)}\left(F, \left(\frac{p_1}{q_1}, \ldots, \frac{p_m}{q_m}\right)\right) < (m+1)\epsilon$.

The statement above is, except for minor notational changes, the statement used in Chapter III.

Proof. We apply the arithmetic version with $n_1 = \cdots = n_m = 1$. Suppose that the index is $\geq (m+1)\epsilon$. Then there exists a decreasing set of irreducible components $Z(a)$ of $Z_{a\epsilon}$ $(a = 1, \ldots, m+1)$ with

$$\mathcal{P} \supset Z(1) \supset Z(2) \supset \cdots \supset Z(m+1) \ni \left(\frac{p_1}{q_1}, \ldots, \frac{p_m}{q_m}\right)$$

Since $\dim(\mathcal{P}) = m$ one has $Z = Z(a) = Z(a+1)$ for some a. The arithmetic version applied to Z yields

$$Z = Z_1 \times \ldots \times Z_m$$

For some i we must have $Z_i = \left\{\frac{p_i}{q_i}\right\}$ and so

$$d_i \log q_i \leq c_1(\epsilon) \log|F| + c_2(\epsilon)(d_1 + \cdots + d_m)$$

hence

$$d_1(\log q_1 - mc_2(\epsilon)) \leq c_1(\epsilon) \log|F|$$

since $d_1 + \cdots + d_m \leq md_1$ and $d_1 \log q_1 \leq d_i \log q_i$. Choose $C_2(\epsilon) = 2mc_2(\epsilon)$. Since $\log q_1 \geq C_2$ one has

$$d_1 \frac{1}{2} \log q_1 \leq c_1(\epsilon) \log|F|$$

Choose $C_1(\epsilon) = 3c_1(\epsilon)$ and we find a contradiction. □

Chapter IX

Geometric Part of Faltings's Proof

by Carel Faber

We follow Faltings's [22], §4.

Let k be an algebraically closed field of characteristic zero, A an abelian variety over k, and $X \subset A$ an irreducible subvariety which does not contain any translate of an abelian subvariety $B \subset A$ of positive dimension.

1 Lemma. *For m big enough the map $\alpha_m: X^m \to A^{m-1}$ defined by $\alpha_m(x_1, \ldots, x_m) = (2x_1 - x_2, 2x_2 - x_3, \ldots, 2x_{m-1} - x_m)$ is finite.*

Proof. We use the following facts:

1. A projective morphism that is quasi-finite (i.e., whose fibres are set-theoretically finite) is finite. ([27], Ch. III, Exc. 11.2.)

2. Let $f: X_1 \to X_2$ be a morphism of projective varieties. Let $\mathcal{L}$ be a line bundle on X_1 that is ample on the fibres of f. Then the set of degrees of fibres of f, measured with respect to $\mathcal{L}$ (see Chapter VII, §4), is finite. Sketch of proof: We may assume that f is surjective. So f is flat on an open subset U of X_1 containing the generic fibre. The image in X_2 of the complement of U is a finite union of irreducible closed subsets of lower dimension, etc. ([27], Ch. III, Exc. 9.4 and Cor. 9.10).

The essential points of the proof are:

1. The equations for the fibre over $\vec{a}_{m-1} := (a_1, \ldots, a_{m-1})$:

$$\begin{cases} x_2 &= 2x_1 - a_1 \\ x_3 &= 2x_2 - a_2 \\ &\vdots \\ x_m &= 2x_{m-1} - a_{m-1} \end{cases}$$

(to be solved with $x_1, x_2, \ldots, x_m$ in X) show that for $m \geq n$ the projection onto the first n factors $p_{m,n}: X^m \to X^n$ induces a closed immersion on fibres $\alpha_m^{-1}(\vec{a}_{m-1}) \hookrightarrow \alpha_n^{-1}(\vec{a}_{n-1})$. In particular, via $p_{m,1}: X^m \to X$, we get closed immersions of fibres of α_m into X.

2. So the maximum of dimensions of fibres of α_m exists, and decreases with m, thus is constant, say equal to d, for $m \geq m_0$.

3. We want to show that $d = 0$: then the morphism α_m has finite fibres, and we are done by Fact 1 above.

4. So suppose that $d > 0$. Fix an ample line bundle $\mathcal{L}$ on A. By the degree of a subvariety Z of a fibre of α_m we will mean the degree of $p_{m,1}Z$ in A, with respect to $\mathcal{L}$. (Note that this is the degree of Z with respect to $p_{m,1}^*\mathcal{L}$.) For all $m \geq m_0$ we look at the d-dimensional irreducible components of fibres of α_m. Any such component is also a component of a fibre of α_{m_0} (by 1. above). Apply Fact 2 above with $\alpha_{m_0}: X^{m_0} \to A^{m_0-1}$ and line bundle $p_{m_0,1}^*\mathcal{L}$. We conclude that for any $m \geq m_0$ the maximum of degrees of d-dimensional irreducible components of fibres of α_m exists, and that it decreases with m. I.e., such degrees are bounded.

5. For $m \geq m_0$ define Y_m to be the subset of A^{m-1} above which the fibres of α_m have dimension d. Since for such m the maximum dimension of a fibre is d, the subsets Y_m are non-empty and closed ([27], Ch. II, Exc. 3.22d). For $m \geq n \geq m_0$, denote by $q_{m,n}: A^{m-1} \to A^{n-1}$ the projection onto the first $n-1$ factors. Then $q_{m,n}(Y_m) \subset Y_n$. The subsets $q_{m,m_0}(Y_m)$ of Y_{m_0} are non-empty and closed, and any finite intersection of them is non-empty. Thus the intersection of all these subsets is non-empty.

6. Pick a point y in this intersection. Among the finitely many d-dimensional components in the fibre above y, there is at least one that occurs in fibres of α_m for infinitely many m; thus (by 1. above) it occurs in fibres of α_m for all m. Let Z be such a d-dimensional component, and consider Z as a subset of A. Note that for any irreducible closed $S \subset A$ we have the following equivalence:

$$S \text{ contained in a fibre of } \alpha_m$$

$$\Longleftrightarrow$$

$$S \subset X \text{ and for all } 0 < r < m \text{ there exists a } b_r \in A \text{ with } 2^r S + b_r \subset X.$$

So for all $r > 0$ there exists a $b_r \in A$ with $2^r Z + b_r \subset X$, and this again implies that $2^r Z + b_r$ is contained in a fibre of α_{m_0}, for all positive r. We conclude (by 4. above) that the degree of $2^r Z$ (which equals the degree of $2^r Z + b_r$) is bounded uniformly in r.

7. Let $G \subset A$ be the algebraic subgroup such that $g \in G$ if and only if $g + Z = Z$. Note that G is closed in A, so the connected component G° of the identity is an abelian subvariety of A. The product of the $\mathcal{L}$-degree of $2^r Z$ and the degree of the map $2^r: Z \to 2^r Z$ equals the $(2^r)^*\mathcal{L}$-degree of Z (see Chapter VII, §4). Since $(2^r)^*\mathcal{L}$ is numerically equivalent to $\mathcal{L}^{\otimes 4^r}$ (Chapter VII, §6.1), the $(2^r)^*\mathcal{L}$-degree of Z equals $(4^r)^d$ times the $\mathcal{L}$-degree of Z. It follows that the degree of the map $2^r: Z \to 2^r Z$, which equals the number of 2^r-torsion points in G, grows like 4rd. We conclude that the dimension of G equals d. By the definition of G we have for any $z \in Z$ that $z + G \subset Z$. So Z contains a translate of a d-dimensional abelian variety (in fact equality holds, and G is an abelian variety).

8. We have shown that α_m is finite for all $m \geq m_0$. It is easy to show that if X is a curve, then one can take $m_0 = 2$. Finally we remark that the assumption is necessary: if $x + B \subset X$ where $B \subset A$ is an abelian subvariety of positive dimension, then for all $b \in B$ and for all m the point $(b + x, 2b + x, 4b + x, \ldots, 2^{m-1}b + x)$ is in the fibre over $(x, x, \ldots, x) \in A^{m-1}$, so α_m has infinite fibres. □

We continue with [22], §4. Choose a big enough integer m such that the map $\alpha_m: X^m \to A^{m-1}$ is finite, and choose a very ample and symmetric line bundle $\mathcal{L}$ on A, embedding $A \subset \mathbb{P}^n$. In the sequel we will mainly use additive notation for the tensor product in the Picard group. E.g., on $A \times A$ we have the Poincaré bundle

$\mathcal{P} = \text{add}^*\mathcal{L} - \text{pr}_1^*\mathcal{L} - \text{pr}_2^*\mathcal{L}$ where add: $A \times A \to A$ denotes the addition. We will also consider linear combinations of line bundles with rational coefficients, i.e., we identify a line bundle $\mathcal{L}$ with its image in $\text{Pic}(\cdot) \otimes \mathbb{Q}$; we can do this since we are interested only in ampleness (see Chapter VII, Remark 5.4).

For $\epsilon, s_1, \ldots, s_m$ positive rational numbers we define on A^m the (rational) line bundle $\mathcal{L}(-\epsilon, s_1, \ldots, s_m)$ as the rational linear combination

$$(1.1) \qquad \mathcal{L}(-\epsilon, s_1, \ldots, s_m) = -\epsilon \cdot \sum_{i=1}^{m} s_i^2 \cdot \text{pr}_i^*(\mathcal{L}) + \sum_{i=1}^{m-1} (s_i x_i - s_{i+1} x_{i+1})^*(\mathcal{L}).$$

Here $(s_i x_i - s_{i+1} x_{i+1})^*(\mathcal{L})$ means

$$\frac{1}{n^2} \cdot (n s_i x_i - n s_{i+1} x_{i+1})^*(\mathcal{L})$$

for any non-zero integer n such that ns_i and ns_{i+1} are integers, so that sending $(x_1, \ldots, x_m)$ in A^m to $ns_i x_i - ns_{i+1} x_{i+1}$ is a morphism. (It follows from Cor. 3.4 that this is well-defined.)

This line bundle $\mathcal{L}(-\epsilon, s_1, \ldots, s_m)$ is a rational linear combination of $\mathcal{L}_i = \text{pr}_i^*(\mathcal{L})$ and $\mathcal{P}_{i,j} = \text{pr}_{i,j}^*(\mathcal{P})$ (use the Theorem of the Cube, Chapter V, Cor. 3.2):

$$\begin{aligned}(1.2) \qquad \mathcal{L}(-\epsilon, s_1, \ldots, s_m) &= (1-\epsilon)s_1^2 \mathcal{L}_1 + \sum_{i=2}^{m-1} (2-\epsilon) s_i^2 \mathcal{L}_i + \\ &\quad + (1-\epsilon) s_m^2 \mathcal{L}_m - \sum_{i=1}^{m-1} s_i s_{i+1} \mathcal{P}_{i,i+1},\end{aligned}$$

so for fixed ϵ the coefficients of $\mathcal{L}_i$ are proportional to s_i^2, and those of $\mathcal{P}_{i,j}$ to $s_i s_j$.

2 Lemma. *Suppose $Y = Y_1 \times \cdots \times Y_m$ is a product subvariety of A^m. Then as a function of the s_i the intersection product $\mathcal{L}(-\epsilon, s_1, \ldots, s_m)^{\dim(Y)} \cdot Y$ is proportional to $\prod_{i=1}^m s_i^{2 \cdot \dim(Y_i)}$.*

Proof. The intersection number is a linear combination of terms $\prod_i \mathcal{L}_i^{e_i} \cdot \prod_{i \neq j} \mathcal{P}_{i,j}^{e_{i,j}} \cdot Y$, with coefficients proportional to $\prod_i s_i^{e_i} \cdot \prod_{i \neq j} (s_i s_j)^{e_{i,j}}$ (see 1.2), and $\sum_i e_i + \sum_{i \neq j} e_{i,j} = \dim(Y)$. We claim that such a term is zero unless for each i one has $2e_i + \sum_{j \neq i} e_{i,j} \leq 2\dim(Y_i)$. To see this, we use the cohomological interpretation of intersection numbers (Chapter VII, §4, Remark g):

$$\prod_i \mathcal{L}_i^{e_i} \cdot \prod_{i \neq j} \mathcal{P}_{i,j}^{e_{i,j}} \cdot Y = \left\langle \wedge_i c_1(\mathcal{L}_i)^{e_i} \bigwedge \wedge_{i \neq j} c_1(\mathcal{P}_{i,j})^{e_{i,j}}, [Y] \right\rangle$$

(note that the order in the wedge product doesn't matter since the factors are of even degree). In terms of de Rham cohomology this means:

$$\prod_i \mathcal{L}_i^{e_i} \cdot \prod_{i \neq j} \mathcal{P}_{i,j}^{e_{i,j}} \cdot Y = \int_{Y^{\text{an}}} \wedge_i c_1(\mathcal{L}_i)^{e_i} \bigwedge \wedge_{i \neq j} c_1(\mathcal{P}_{i,j})^{e_{i,j}}$$

where c_1 now denotes the first Chern class in $H^2_{\text{DR}}((A^m)^{\text{an}})$ (to integrate a volume form on a possibly singular variety like Y^{an}, choose a finite projection $Y \to \mathbb{P}^{\dim Y}$ as in Lemma 5 below and perform the integration on $(\mathbb{P}^{\dim Y})^{\text{an}}$). Note that for each

i, $c_1(\mathcal{L}_i)^{e_i} = p_i^*(c_1(\mathcal{L})^{e_i})$ is a pullback of a differential form of degree $2e_i$ on Y_i. Now think of A^{an} as $\mathbb{C}^{\dim A}$ modulo a lattice. Then add: $A^{\text{an}} \times A^{\text{an}} \to A^{\text{an}}$ is induced by $+\colon \mathbb{C}^{\dim A} \times \mathbb{C}^{\dim A} \to \mathbb{C}^{\dim A}$ and $c_1(\mathcal{L})$ is represented by a linear combination of terms $dx \wedge dy$ with x, y in $\mathrm{Hom}_{\mathbb{R}}(\mathbb{C}^{\dim A}, \mathbb{R})$. The calculation:

$$\begin{aligned} &\mathrm{add}^*(dx \wedge dy) - p_1^*(dx \wedge dy) - p_2^*(dx \wedge dy) = \\ &= d(x_1 + x_2) \wedge d(y_1 + y_2) - dx_1 \wedge dy_1 - dx_2 \wedge dy_2 \\ &= dx_1 \wedge dy_2 + dx_2 \wedge dy_1 \end{aligned}$$

where $x_i = p_i^* x$ and $y_i = p_i^* y$, shows that $c_1(\mathcal{P}_{i,j})$ is a linear combination of terms $p_i^*\omega_i \wedge p_j^*\omega_j$ with ω_i, ω_j in $H^1_{\mathrm{DR}}(A^{\text{an}})$. The claim follows.

As $\sum_i (2e_i + \sum_{j \neq i} e_{i,j}) = \sum_i 2\dim(Y_i)$, the intersection number is zero unless $2e_i + \sum_{j\neq i} e_{i,j} = 2\dim(Y_i)$ for all i. □

3 Corollary. *There exists a positive ϵ_0 such that for any $\epsilon \leq \epsilon_0$ and for any product variety $Y \subset X^m$ the intersection number $\mathcal{L}(-\epsilon, s_1, \ldots, s_m)^{\dim(Y)} \cdot Y$ is positive.*

Proof. Indeed, for $s_i = 2^{m-i}$ we have that $\mathcal{L}(0, s_1, \ldots, s_m)$ is the pull-back by the *finite* morphism α_m of the ample line bundle $4^{m-2}\mathcal{L}_1 + 4^{m-3}\mathcal{L}_2 + \cdots + \mathcal{L}_{m-1}$, hence it is ample on X^m (see Chapter VII, Cor. 3.2). Hence for these s_i and for small ϵ the bundle $\mathcal{L}(-\epsilon, s_1, \ldots, s_m)$ is ample, since the ample cone is open (see Chapter VII, Thm. 4.3.1). It follows that the implied constant in Lemma 2 (depending on ϵ and Y) is positive for small ϵ and arbitrary Y, showing that the intersection number $\mathcal{L}(-\epsilon, s_1, \ldots, s_m)^{\dim(Y)} \cdot Y$ is positive for small ϵ, arbitrary Y and arbitrary (positive) $s_1, \ldots, s_m$. □

The main result of this section can now be formulated.

4 Theorem. *Let A be an abelian variety over an algebraically closed field k of characteristic zero. Let $X \subset A$ be an irreducible subvariety which does not contain any translate of a positive dimensional abelian subvariety of A. Take m big enough as in Lemma 1 and let $\mathcal{L}$ be a symmetric ample line bundle on A. For ϵ, $s_1, \ldots, s_m$ positive rational numbers let $\mathcal{L}(-\epsilon, s_1, \ldots, s_m)$ be the element of $\mathrm{Pic}(A^m) \otimes \mathbb{Q}$ given by 1.1. Take ϵ_0 as in Cor. 3. For any $\epsilon < \epsilon_0$ there exists an integer s, such that $\mathcal{L}(-\epsilon, s_1, \ldots, s_m)$ is ample on X^m if $s_1/s_2 \geq s$, $s_2/s_3 \geq s$, $\ldots$, $s_{m-1}/s_m \geq s$.*

Before proving this result, we need some information concerning projections. Let $L \subset \mathbb{P}^n$ be a linear subvariety of dimension m. After a suitable choice of homogeneous coordinates $x_0, \ldots, x_n$, one can assume that $L = V(x_{m+1}, \ldots, x_n)$. Then L gives rise to a projection morphism $\pi\colon \mathbb{P}^n - L \to \mathbb{P}^{n-m-1}$, sending $(a_0 \colon \cdots \colon a_n)$ to $(a_{m+1} \colon \cdots \colon a_n)$. Let $(a_{m+1} \colon \cdots \colon a_n) \in \mathbb{P}^{n-m-1}$, then the map $\mathbb{A}^{m+1} \to \mathbb{P}^n$ sending $(a_0, \ldots, a_m)$ to $(a_0 \colon \cdots \colon a_n)$ is an isomorphism between $\mathbb{A}^{m+1}$ and $\pi^{-1}(a_{m+1} \colon \cdots \colon a_n)$; hence the fibres of π are affine. Suppose now that $X \subset \mathbb{P}^n$ is an irreducible subvariety of dimension d, such that $X \cap L = \emptyset$. Then $\pi\colon X \to \pi X$ is finite (since it is projective and its fibres are affine), and $\deg(X) = \deg(\pi\colon X \to \pi X) \cdot \deg(\pi X)$ (see Chapter VII, §4).

5 Lemma. *Let $X \subset \mathbb{P}^n$ be an irreducible subvariety of dimension d. There exists a finite projection $\pi\colon X \to \mathbb{P}^d$ of degree $\deg(X)$ and a global section $s \neq 0$ in $\Gamma(\mathbb{P}^n, \mathcal{O}(N))$ for some $N \leq (n-d)\deg(X)$, such that the ideal sheaf of $V(s)$ annihilates $\Omega^1_{X/\mathbb{P}^d}$. Especially, π is étale on $X - (X \cap V(s))$. Moreover, $\mathrm{Norm}_\pi(s|_X)$ in $\Gamma(\mathbb{P}^d, \mathcal{O}(N\deg(X)))$ defines a hypersurface whose ideal sheaf annihilates $\Omega^1_{X/\mathbb{P}^d}$.*

Proof. Let $L_0 \subset \mathbb{P}^n$ be a linear subvariety of codimension $d+2$, such that $L_0 \cap X = \emptyset$. This gives a projection $\pi_0: \mathbb{P}^n - L_0 \to \mathbb{P}^{d+1}$ and $\pi_0 X \subset \mathbb{P}^{d+1}$ is a hypersurface of degree $\leq \deg(X)$. It follows that $X \subset \pi_0^{-1}\pi_0 X = V(F)$ with F an irreducible homogeneous polynomial of degree $\leq \deg(X)$. Let $x_0, \ldots, x_n$ be homogeneous coordinates on $\mathbb{P}^n$ such that $\partial F/\partial x_n \neq 0$ and $P := (0:\cdots:0:1) \notin V(F)$. Let $\pi_1: \mathbb{P}^n - \{P\} \to \mathbb{P}^{n-1}$ be the projection given by P. A computation shows that the ideal sheaf of $V(\partial F/\partial x_n)$ is the annihilator of $\Omega^1_{V(F)/\mathbb{P}^{n-1}}$. By induction we can assume that the lemma has been proved for $\pi_1 X \subset \mathbb{P}^{n-1}$; let $\pi_2: \pi_1 X \to \mathbb{P}^d$ and $s_1 \in \Gamma(\mathbb{P}^{n-1}, \mathcal{O}(N_1))$ be as required. Then $\pi := \pi_2 \circ \pi_1$ and $s := (\partial F/\partial x_n)\pi_1^* s_1$ satisfy our requirements. For the construction, definition and properties of $\mathrm{Norm}_\pi(s|_X)$ see [25], II, §6.5. Note that in our case π is not necessarily finite locally free. $\square$

We will also need the following lemma, the proof of which was communicated to me by Edixhoven.

6 Lemma. *Let X be a projective variety, let D be an ample effective Cartier divisor on X and let $\mathcal{M}$ be an invertible sheaf on X such that the restriction of $\mathcal{M}$ to D is ample. Then for d large enough we have $h^i(X, \mathcal{M}^{\otimes d}) = 0$ for all $i \geq 2$.*

Proof. We claim that for d big enough we have that for all $i \geq 1$ and all $n \geq 0$ the cohomology groups $H^i(D, \mathcal{M}^{\otimes d}(nD)|_D)$ vanish. Assume this for a moment, and take d as in the claim. Also, take n big enough such that $h^i(X, \mathcal{M}^{\otimes d}(nD)) = 0$ for all $i \geq 1$ (note D is ample, cf. [27], Ch. III, Prop. 5.3).

We consider thickenings of D: let D_n be the n-th infinitesimal neighbourhood of D in X, i.e., the closed subscheme with ideal sheaf $\mathcal{I}_D^{n+1}$, where $\mathcal{I}_D$ is the ideal sheaf of D in X. Then the long exact cohomology sequence belonging to the standard short exact sequence

$$0 \to \mathcal{M}^{\otimes d} \to \mathcal{M}^{\otimes d}(nD) \to \mathcal{M}^{\otimes d}(nD)|_{D_{n-1}} \to 0$$

gives isomorphisms

$$H^i(D_{n-1}, \mathcal{M}^{\otimes d}(nD)|_{D_{n-1}}) \simeq H^{i+1}(X, \mathcal{M}^{\otimes d})$$

for all $i \geq 1$. So we want to show that for $i \geq 1$ we have $H^i(D_{n-1}, \mathcal{M}^{\otimes d}(nD)|_{D_{n-1}}) = 0$. For this we use the filtration of coherent sheaves on D_{n-1}:

$$\mathcal{M}^{\otimes d}(nD)|_{D_{n-1}} \supset \mathcal{M}^{\otimes d}(nD)\cdot\mathcal{I}_D|_{D_{n-1}} \supset \cdots \supset \mathcal{M}^{\otimes d}(nD)\cdot\mathcal{I}_D^{n-1}|_{D_{n-1}} \supset 0$$

whose successive quotients are

$$\mathcal{M}^{\otimes d}(nD)|_D,\ \mathcal{M}^{\otimes d}((n-1)D)|_D,\ \ldots, \mathcal{M}^{\otimes d}(D)|_D\,.$$

From the short exact sequences associated to the filtration we see that indeed

$$H^i(D_{n-1}, \mathcal{M}^{\otimes d}(nD)|_{D_{n-1}}) = 0$$

for all $i \geq 1$.

It remains to prove the claim. We prove a more general statement. Let X be a projective variety, $\mathcal{M}$ and $\mathcal{N}$ ample line bundles on X and $\mathcal{F}$ a coherent sheaf on X. Let $\mathcal{F}(a,b) := \mathcal{F} \otimes \mathcal{M}^{\otimes a} \otimes \mathcal{N}^{\otimes b}$. Then the set

$$S_{\mathcal{F}} := \{(a,b) \mid a \geq 0,\ b \geq 0,\ h^i(X, \mathcal{F}(a,b) \neq 0 \text{ for some } i > 0\}$$

is finite.

We may suppose that $\mathcal{M}$ and $\mathcal{N}$ are very ample (replace $\mathcal{M}$ by $\mathcal{M}^m$, $\mathcal{N}$ by $\mathcal{N}^n$, $\mathcal{F}$ by the direct sum $\oplus \mathcal{F}(a,b)$ with $0 \leq a < m$, $0 \leq b < n$). Then we have an embedding of X in a product of two projective spaces. Now we mimick the argument for one projective space in [27], Ch. III, proof of Thm. 5.2: we may replace X by the product of the two projective spaces, and $\mathcal{F}$ by any coherent sheaf on that product. There is a short exact sequence

$$0 \to \mathcal{R} \to \mathcal{E} \to \mathcal{F} \to 0$$

with $\mathcal{E}$ a finite direct sum of sheaves $\mathcal{O}(a_i, b_i)$, and $\mathcal{R}$ coherent. All we need now is that $\mathcal{E}(j,k)$ has no higher cohomology for j and k big enough. This follows from the corresponding statement for $\mathcal{O}$; for this, use a Künneth formula as in [71], Ch. VII, Prop. 12 and Remarque, pp. 185–186 or in [25], Ch. III, Thm. 6.7.8, or, alternatively, use the Kodaira vanishing theorem. This finishes the proof of the claim and the lemma. □

We can now begin the proof of Thm. 4.

Proof. Let $\epsilon < \epsilon_0$. By induction on r we will show:

> For any integers N and r there exists an integer s_0, such that for any product subvariety $Y = Y_1 \times \cdots \times Y_m$ of X^m with $\dim(Y) = r$ and $\deg(Y_i) \leq N$ for all i, $\mathcal{L}(-\epsilon, s_1, \ldots, s_m)$ is ample on Y if $s_1/s_2 \geq s_0$, $\ldots, s_{m-1}/s_m \geq s_0$.

For $r = 0$ this is trivially true, and for $r = m \cdot \dim(X)$ and $N = \deg(X)$ we get the statement of the theorem. So assume the statement is proven for all $r < r_0$, and let's try to prove it for $r = r_0$. Let N be given.

First of all, if we take the ratios $s_1/s_2, \ldots, s_{m-1}/s_m$ big enough, then for all $Y = Y_1 \times \cdots \times Y_m \subset X^m$ with $\dim(Y) = r_0$ and $\deg(Y_i) \leq N$, there exists an effective ample divisor of Y such that the restriction of $\mathcal{L}(-\epsilon, s_1, \ldots, s_m)$ to that divisor is ample. Namely, take $D_1 + \cdots + D_m$, with $D_i = H_i \times \prod_{j \neq i} Y_j$, $H_i \subset Y_i$ a hyperplane section; apply the induction hypothesis with $r = r_0 - 1$, same N. This will be used later on.

We will show that, if $s_1/s_2, \ldots, s_{m-1}/s_m$ are sufficiently big, and Y is as in the statement, $\mathcal{L}(-\epsilon, s_1, \ldots, s_m)$ has non-negative degree on all irreducible curves $C \subset Y$. By Kleiman's theorem (cf. Chapter VII, Thm. 5.5) it follows that $\mathcal{L}(-\epsilon, s_1, \ldots, s_m)$ is in the closure of the ample cone. Decreasing ϵ a little bit then gives an ample line bundle (note that we could have started with $(\epsilon + \epsilon_0)/2$ instead of ϵ).

So let Y be as in the statement, and $C \subset Y$ an irreducible curve. The idea of the proof is now as follows. One shows that if $s_1/s_2, \ldots, s_{m-1}/s_m$ are sufficiently big, for d sufficiently big and divisible there exists a non-trivial global section f of $\mathcal{L}(-\epsilon, s_1, \ldots, s_m)^{\otimes d}$ on Y. If the restriction of f to C is non-zero, then of course $\mathcal{L}(-\epsilon, s_1, \ldots, s_m)$ has non-negative degree on C. If the restriction of f to C is zero, one distinguishes two cases: f has small or large index along C. If f has large index along C, one uses the Product Theorem to show that C is contained in a product variety of dimension less than r_0 (and of bounded degree), and one uses induction. If f has small index along C, then one constructs a derivative of f that does not vanish on C and has suitably bounded poles. How to define these two cases and how big $s_1/s_2, \ldots, s_{m-1}/s_m$ should be will follow from the computations below. Of course all estimates concerning the s_i/s_{i+1} will have to be done uniformly in Y and in C.

As Faltings puts it, "set up the geometry": Choose projections $\pi_i: Y_i \to \mathbb{P}^{n_i} = P_i$ (with $\deg(\pi_i) = \deg(Y_i)$) and (not necessarily reduced) hypersurfaces $Z_i \subset P_i$ of degree $\deg(Z_i) \le nN^2$, such that the ideal sheaf of Z_i annihilates $\Omega^1_{Y_i/P_i}$ (cf. Lemma 5; the n comes from the $\mathbb{P}^n$ in which A is embedded). Let $\pi: Y \to P = P_1 \times \cdots \times P_m$ be the product of the π_i; then $\deg(\pi)$ is the product of the $\deg(Y_i)$. We claim that any derivation ∂ on P_i (there are many because P_i is a homogeneous space under $\mathrm{GL}(n_i{+}1)$) extends to a derivation on Y_i with at most a simple pole along $\pi_i^* Z_i$. To see this, consider the exact sequence

$$0 \to \pi_i^* \Omega^1_{P_i/k} \to \Omega^1_{Y_i/k} \to \Omega^1_{Y_i/P_i} \to 0$$

of [27], Ch. II, Prop. 8.11. Applying $\mathrm{Hom}_{\mathcal{O}_{Y_i}}(\cdot, \pi_i^*\mathcal{O}(\deg(Z_i)))$ to it gives the exact sequence

$$\begin{aligned} 0 \to \mathrm{Der}_k(\mathcal{O}_{Y_i}, \pi_i^*\mathcal{O}(\deg(Z_i))) &\to \mathrm{Hom}_{\mathcal{O}_{Y_i}}(\pi_i^*\Omega^1_{P_i/k}, \pi_i^*\mathcal{O}(\deg(Z_i))) \to \\ &\to \mathrm{Ext}^1_{\mathcal{O}_{Y_i}}(\Omega^1_{Y_i/P_i}, \pi_i^*\mathcal{O}(\deg(Z_i))) \end{aligned}$$

Let G_i in $\Gamma(P_i, \mathcal{O}(\deg(Z_i)))$ be an equation for Z_i. Saying that the ideal sheaf of Z_i annihilates $\Omega^1_{Y_i/P_i}$ means that the map

$$\cdot G_i: \Omega^1_{Y_i/P_i} \to \pi_i^*\mathcal{O}(\deg(Z_i)) \otimes_{\mathcal{O}_{Y_i}} \Omega^1_{Y_i/P_i}$$

is zero. One checks that $\pi_i^*(G_i\partial)$ maps to zero in the Ext^1.

If C is contained in the preimage of some Z_i, i.e.,

$$C \subset \pi_i^{-1} Z_i \times \prod_{j \ne i} Y_j$$

then we are done by induction: note that $\pi_i: \pi_i^{-1} Z_i \to Z_i$ is finite of degree $\deg(Y_i) \le N$ (since $\pi_i: Y_i \to P_i$ is flat in codimension one), hence $\deg(\pi_i^{-1} Z_i) \le N \deg(Z_i) \le nN^3$ by Chapter VII, §4, Remark (f). So suppose that C is not contained in the inverse image of any of the Z_i. Then $\pi: C \to D = \pi C$ is generically étale.

For $s_1/s_2, \ldots, s_{m-1}/s_m$ big enough, and d big enough and sufficiently divisible, we have that $\Gamma(Y, \mathcal{L}(-\epsilon, s_1, \ldots, s_m)^{\otimes d})$ is non-trivial. This follows from:
1. For such d the Euler characteristic $\chi(Y, \mathcal{L}(-\epsilon, s_1, \ldots, s_m)^{\otimes d})$ is positive.
2. For such d and $i \ge 2$ the cohomology groups $H^i(Y, \mathcal{L}(-\epsilon, s_1, \ldots, s_m)^{\otimes d})$ vanish.

The second statement is a direct consequence of Lemma 6. To get the first statement, recall (Chapter VII, §4) that, as a function of d, $\chi(Y, \mathcal{L}(-\epsilon, s_1, \ldots, s_m)^{\otimes d})$ is a polynomial of degree $\le \dim(Y)$, and that almost by definition the coefficient of $d^{\dim(Y)}$ is $1/\dim(Y)!$ times the intersection number $(\mathcal{L}(-\epsilon, s_1, \ldots, s_m)^{\dim(Y)}.Y)$, which is positive by Cor. 3 and the choice of ϵ_0. So, if $s_1/s_2, \ldots, s_{m-1}/s_m$ are big enough (uniformly in Y and C), then for d sufficiently big and divisible, $\mathcal{L}(-\varepsilon, s_1, \ldots, s_m)^{\otimes d}$ has a non-trivial global section f on Y.

Next we consider line bundles on A^m. We have the identity in $\mathrm{Pic}(A^m)\otimes\mathbb{Q}$ (use the Theorem of the Cube and that $\mathcal{L}$ is symmetric):

$$(s_ix_i - s_{i+1}x_{i+1})^*(\mathcal{L}) + (s_ix_i + s_{i+1}x_{i+1})^*(\mathcal{L}) = 2s_i^2\mathcal{L}_i + 2s_{i+1}^2\mathcal{L}_{i+1}$$

Suitable multiples of the two terms on the left are generated by their global sections (because $\mathcal{L}$ is). Hence we can find sections without common zeroes. Multiplying by these, we first find "injections without common zeroes" on A^m:

$$d\cdot\sum_{i=1}^{m-1}(s_ix_i - s_{i+1}x_{i+1})^*(\mathcal{L}) \to d\cdot(2s_1^2\mathcal{L}_1 + 4s_2^2\mathcal{L}_2 + \cdots + 4s_{m-1}^2\mathcal{L}_{m-1} + 2s_m^2\mathcal{L}_m)$$

then, using that the $\mathcal{L}_i$ are also generated by global sections, we find injections without common zeroes

$$\begin{aligned} d\cdot\mathcal{L}(-\epsilon, s_1,\ldots,s_m) &\to d\cdot((2-\epsilon)s_1^2\mathcal{L}_1 + (4-\epsilon)s_2^2\mathcal{L}_2 + \cdots + (2-\epsilon)s_m^2\mathcal{L}_m) \\ &\to d\cdot(4s_1^2\mathcal{L}_1 + \cdots + 4s_m^2\mathcal{L}_m) \end{aligned}$$

Define $d_i = 4ds_i^2$, and choose an injection $\rho\colon \mathcal{L}(-\varepsilon, s_1,\ldots,s_m)^{\otimes d} \to \otimes_{i=1}^m \mathcal{L}_i^{d_i}$ which does not vanish identically on C (i.e., ρ is an isomorphism at the generic point of C).

We come to the final part of the proof. Define the index $i(C,f)$ of f along C to be the index $i(x,f)$ of f at the generic point of C, with respect to the weights d_i (cf. Chapter VIII, Def. 1.8). Since ρ is an isomorphism at x, we have $i(x,f) = i(x,\rho(f))$. Now note that $\mathcal{L}_i = \pi_i^*\mathcal{O}(1)$ since π_i is a projection from $\mathbb{P}^n$ minus a linear subvariety to P_i and A was embedded in $\mathbb{P}^n$ via global sections of $\mathcal{L}$. So $\rho(f)$ can be viewed as a section on Y of $\pi^*\mathcal{O}(d_1,\ldots,d_m)$. Because $\pi\colon Y \to P$ is finite and surjective, and Y is integral and P is integral and normal, we can take the norm of f with respect to π (cf. [25], II, §6.5):

$$g := \mathrm{Norm}_\pi(\rho(f)) \in \Gamma(P, \mathcal{O}(d_1,\ldots,d_m)^{\otimes\deg(\pi)}) = \Gamma(P, \mathcal{O}(\deg(\pi)d_1,\ldots,\deg(\pi)d_m)$$

Let $i(D,g) := i(\pi(x),g)$ be the index of g at $\pi(x)$ (i.e., along $D=\pi C$) with respect to the weights $\deg(\pi)d_i$. We claim that $i(D,g) \geq i(C,f)/\deg(\pi)$. To see this, note that π is étale above $\pi(x)$, so that over some étale neighbourhood U of $\pi(x)$, $\pi\colon Y \to P$ is a disjoint union of $\deg(\pi)$ copies of U; then use the formula $i(x,f_1f_2) = i(x,f_1)+i(x,f_2)$.

Now we choose a sufficiently small positive number σ (the precise choice of σ will be explained later), and we distinguish two cases.

Case 1: $i(D,g) \geq \sigma$. Then apply the Product Theorem (Chapter VIII, Thm. 2.1) as in Chapter VIII, Remark 2.3. To be precise: there exists a chain

$$P \neq Z_1 \supset Z_2 \supset \cdots \supset Z_{\dim(P)+1} \ni x$$

with Z_l an irreducible component of the closed subscheme of P where g has index $\geq l\sigma/(\dim(P)+1)$. Because of dimensions, there exists an l with $1 \leq l \leq \dim(P)$, such that $Z_l = Z_{l+1}$. The Product Theorem then shows that if the s_i/s_{i+1} are sufficiently big (in terms of σ, m, and $n_1,\ldots,n_m$), then $Z_l = V_1\times\cdots\times V_m$ is a product variety of dimension $< \dim(P)$ (since $f\neq 0$) and the $\deg(V_i)$ are bounded in terms of σ, m, and $n_1,\ldots,n_m$. Let Y' be an irreducible component of $\pi^{-1}Z_l$ with $C\subset Y'$. Then $\dim(Y') < \dim(Y)$, $Y' = Y_1'\times\cdots\times Y_m'$ is a product variety and $\deg(Y_i') \leq \deg(\pi_i)\deg(V_i) \leq N\deg(V_i)$ (note that Y_i' is an irreducible component of $\pi_i^{-1}V_i$ and use Chapter VII, Remark 4.3.f). By induction, if the s_i/s_{i+1} are sufficiently big in terms of σ, m, $n_1,\ldots,n_m$ and N, then $\mathcal{L}(-\varepsilon,s_1,\ldots,s_m)$ is ample on Y', hence has positive degree on C.

Case 2: $i(D,g)<\sigma$. Then $i(C,f)<\sigma\deg(\pi)$. For $1\leq i\leq m$ let $\partial_{i,j}$, $1\leq j\leq n_i$, be global vector fields on P_i which give a basis of the stalk of $\mathbf{Hom}_{\mathcal{O}_{P_i}}(\Omega^1_{P_i/k},\mathcal{O}_{P_i})$ at $\pi_i(x)$. As $\pi\colon Y\to P$ is étale at x, we may view the $\partial_{i,j}$ as derivations of $\mathcal{O}_{Y,x}$. Let f_1 be a generator of the stalk $\mathcal{L}(-\varepsilon,s_1,\ldots,s_m)^{\otimes d}_x$; then $f=hf_1$ for a unique $h\in\mathcal{O}_{Y,x}$. By the definition of index (Chapter VIII, Def. 1.8, also recall what we mean by the value at x of some $f\in\mathcal{O}_{Y,x}$), there exist integers $e_{i,j}\geq 0$ with $\sum_{i,j}e_{i,j}=i(x,f)$, such that $(H(h))(x)\neq 0$, where $H=\prod_{i,j}\partial_{i,j}^{e_{i,j}}$ (in some order). We want to define $(H(f))(x)$ as

$$(H(h))(x)\cdot f_1(x)\in\mathcal{L}(-\varepsilon,s_1,\ldots,s_m)^{\otimes d}(x)=\mathcal{L}(-\varepsilon,s_1,\ldots,s_m)^{\otimes d}_x\otimes_{\mathcal{O}_{Y,x}}k(x)$$

so let us check that this is well-defined. Let $u\in\mathcal{O}^*_{Y,x}$ and $f_1':=u^{-1}f_1$. Then $h'=uh$ and $H(h')=H(uh)=uH(h)+$ terms like $H_1(u)H_2(h)$ with H_2 of lower differential degree than H, hence $(H_2(h))(x)=0$ by the definition of $i(x,f)$. It follows that $(H(f))(x)$ is well-defined.

Since $k(x)$ is the function field of C, $(H(f))(x)$ is a rational section on C of the line bundle $\mathcal{L}(-\varepsilon,s_1,\ldots,s_m)^{\otimes d}$; we want to bound its poles. We have already remarked that each $\partial_{i,j}$ extends to a derivation on Y_i with at most a simple pole along $\pi_i^*Z_i$. Working on an affine open on which $\mathcal{L}(-\varepsilon,s_1,\ldots,s_m)^{\otimes d}$ has a generator f_1 we can write $f=hf_1$ and $\partial_{i,j}=g_i^{-1}\tilde{\partial}_{i,j}$ where the $\tilde{\partial}_{i,j}$ are regular derivations and g_i is an equation for $\pi_i^*Z_i$. Then

$$H=\prod_{i,j}\partial_{i,j}^{e_{i,j}}=\prod_{i,j}(g_i^{-1}\tilde{\partial}_{i,j})^{e_{i,j}}=(\prod_{i,j}g_i^{-e_{i,j}})(\prod_{i,j}\tilde{\partial}_{i,j}^{e_{i,j}})+\cdots$$

where the dots stand for terms of lower differential degree. It follows that

$$(H(h))(x)=\left((\prod_{i,j}g_i^{-e_{i,j}})(\prod_{i,j}\tilde{\partial}_{i,j}^{e_{i,j}})(h)\right)(x)$$

hence its poles are no worse than those of $\prod_{i,j}g_i^{-e_{i,j}}=\prod_i g_i^{-e_i}$, with $e_i=\sum_j e_{i,j}$. Since Y can be covered by affine opens on which $\mathcal{L}(-\varepsilon,s_1,\ldots,s_m)^{\otimes d}$ is trivial, we get a non-zero global section $\prod_{i,j}G_i^{e_{i,j}}(H(f))(x)$ on C of

$$\bigotimes_{i=1}^m\mathcal{L}_i^{\otimes e_i\deg(G_i)}\otimes\mathcal{L}(-\varepsilon,s_1,\ldots,s_m)^{\otimes d}=\bigotimes_{i=1}^m\mathcal{L}_i^{\otimes(e_i\deg(G_i)-\epsilon ds_i^2)}\otimes\bigotimes_{i=1}^{m-1}(s_ix_i-s_{i+1}x_{i+1})^*\mathcal{L}^{\otimes d}$$

where, as before, G_i is an equation for Z_i in $\Gamma(P_i,\mathcal{O}(\deg(Z_i)))$. Using $e_i/ds_i^2=4e_i/d_i\leq i(C,f)<\sigma\deg(\pi)$ and $\deg(G_i)\leq nN^2$ one gets a non-zero global section on C of $\mathcal{L}(4nN^2\deg(\pi)\sigma-\epsilon,s_1,\ldots,s_m)^{\otimes d}$, so this line bundle has non-negative degree on C.

We still have to choose the right σ: we take $\sigma>0$ such that

$$\epsilon':=4nN^2\deg(\pi)\sigma+\epsilon<\epsilon_0$$

The argument above, with ϵ replaced by ϵ', then shows that $\mathcal{L}(-\varepsilon,s_1,\ldots,s_m)$ has non-negative degree on C. □

Acknowledgement. I would like to thank Bas Edixhoven for his help in preparing these notes.

Chapter X

Faltings's Version of Siegel's Lemma

by Robert-Jan Kooman

Siegel's Lemma in its original form guarantees the existence of a small non-trivial integral solution of a system of linear equations with rational integer coefficients and with more variables than equations. It reads as follows:

1 Lemma. (C.L. Siegel) *Let $A = (a_{ij})$ be an $N \times M$ matrix with rational integer coefficients. Put $a = \max_{i,j} |a_{ij}|$. Then, if $N < M$, the equation $Ax = 0$ has a solution $x \in \mathbb{Z}^M, x \neq 0$, with*

$$\|x\| \leq (Ma)^{N/(M-N)}$$

where $\| \ \|$ denotes the max-norm: $\|x\| = \|(x_1, \ldots, x_M)\| = \max_{1 \leq i \leq M} |x_i|$ in $\mathbb{R}^M$.

Stated in an alternative form, the lemma finds a non-trivial lattice element of small norm which lies in the kernel of a linear map α from $\mathbb{R}^M$ to $\mathbb{R}^N$ with the $\mathbb{Z}$-lattices $\mathbb{Z}^M$ and $\mathbb{Z}^N$ where $M > N$ and $\alpha(\mathbb{Z}^M) \subset \mathbb{Z}^N$. Faltings [22] uses a more general variant of the lemma, firstly by taking a general $\mathbb{Z}$-lattice in a normed $\mathbb{R}$-vector space, and secondly, by looking not for one, but for an arbitrary number of linearly independent lattice elements that lie in the kernel of α and are not zero. The lemma is a corollary of Minkowski's Theorem, which we state after the following definitions:

2 Definitions. For V a finite dimensional normed real vector space with $\mathbb{Z}$-lattice M (i.e., M is a discrete subgroup of V that spans V) we define $\lambda_i(V, M)$ as the smallest number λ such that in M there exist i linearly independent vectors of norm not exceeding λ. Further, by V/M we denote a set

$$\{v \in V : v = \lambda_1 v_1 + \ldots + \lambda_b v_b, \ \ 0 \leq \lambda_i < 1 \text{ for } i = 1, \ldots, b = \dim(V)\}$$

where $v_1, \ldots, v_b$ is a basis of M. Finally, if V, W are normed vector spaces with norms $\|\cdot\|_V$, $\|\cdot\|_W$, then $B(V)$ denotes the unit ball $\{x \in V : \|x\|_V \leq 1\}$ in V and if α is a linear map from V to W, the norm $\|\alpha\|$ of α is defined as the supremum of the $\|\alpha(x)\|_W / \|x\|_V$ with $x \in V, \ x \neq 0$. □

We can endow V with a Lebesgue measure μ_V as follows. Take any isomorphism of $\mathbb{R}$-vector spaces $\psi : V \to \mathbb{R}^b$ and let λ denote the Lebesgue measure on $\mathbb{R}^b$. Then, for Lebesgue measurable $A \subset \mathbb{R}^b$, put

$$\mu_V(\psi^{-1}(A)) = \lambda(A).$$

Up to a constant, there is only one Lebesgue measure on V. Hence the quantity

$$\mathrm{Vol}(V) = \mathrm{Vol}(V, \|\cdot\|, M) := \frac{\mu_V(B(V))}{\mu_V(V/M)}$$

does not depend on the choice of μ_V. Clearly, it is also independent of the choice of the basis of M in the definition of V/M.

3 Theorem. (H. Minkowski) *Let V be a normed real vector space of finite dimension b and with $\mathbb{Z}$-lattice M. Then*

$$2^b/b! \leq \lambda_1(V,M)\cdots\lambda_b(V,M)\cdot \mathrm{Vol}(V) \leq 2^b.$$

Proof. See [48]. Note that $B(V)$ is a convex closed body in V, symmetric with respect to the origin, and that $\lambda_1(V,M),\ldots,\lambda_b(V,M)$ are precisely the successive minima of $B(V)$ with respect to M. □

We now state Faltings's version of Siegel's Lemma.

4 Lemma. (Lemma 1 of [22]) *Assume that we have two normed real vector spaces V and W with $\mathbb{Z}$-lattices M and N, respectively, and a linear map $\alpha : V \to W$ such that $\alpha(M) \subset N$. Let $C \geq 2$ be a real number such that the map α has norm at most C, that M is generated by elements of norm at most C and that every non-zero element of M and N has norm at least C^{-1}. Then, for $a = \dim(\ker(\alpha))$, $b = \dim(V)$, and $U = \ker(\alpha)$ with the induced norm on $\ker(\alpha)$,*

$$\lambda_{i+1}(U, U\cap M) \leq (C^{3b}\cdot b!)^{1/(a-i)}$$

for $0 \leq i \leq a-1$.

Thus, for any subset Y of $\ker(\alpha)$ of dimension not exceeding i we can find an element in $\ker(\alpha)$ which is not in Y and whose norm is bounded by the right-hand side of the above equation. The lemma will be applied in situations where the dimensions a and b tend to infinity and such that $b/(a-i)$ remains bounded. Then the right-hand side of the equation grows like a fixed power of C times a power of b.

For the proof of the lemma we let $\|\cdot\|$ be the norm on V, and $\|\cdot\|_U$ its restriction to U. We endow $\alpha(V)$ with the quotient norm: for $v^* \in \alpha(V)$ we put

$$\|v^*\|^* = \inf\{\|v\| : \alpha(v) = v^*\}.$$

The unit balls $B(U)$ and $B(\alpha(V))$ are now defined in a similar way as $B(V)$ and $\mathrm{Vol}(U) = \mathrm{Vol}(U, \|\cdot\|_U, U\cap M)$, $\mathrm{Vol}(\alpha(V)) = \mathrm{Vol}(\alpha(V), \|\cdot\|^*, \alpha(M))$. We first prove the following lemma.

5 Lemma. $\mathrm{Vol}(V) \leq 2^a\cdot\mathrm{Vol}(U)\cdot\mathrm{Vol}(\alpha(V))$.

Proof. Let μ_V, μ_U be Lebesgue measures on V and U, respectively. On $\alpha(V)$ we have a unique Lebesgue measure $\mu_{\alpha(V)}$, defined by

$$\mu_V(E) = \int_{\alpha(V)} f_E(v^*)\, d\mu_{\alpha(V)}(v^*)$$

for μ_V-measurable sets $E \subset V$ where for $v^* = \alpha(v) \in \alpha(V)$ we have

$$f_E(v^*) = \mu_U(\{u \in U : u + v \in E\})$$

which is independent of v since μ_U is translation invariant. We compute $f_{B(V)}(v^*)$ for $v^* \in \alpha(V)$. If $v^* \notin B(\alpha(V))$, then $\|v\| > 1$ so $v \notin B(V)$ for all $v \in \alpha^{-1}(v^*)$ and $f_{B(V)}(v^*) = 0$. If $v^* \in B(\alpha(V))$, then $v \in B(V)$ for some $v \in \alpha^{-1}(v^*)$. For $u \in U$ with $u + v \in B(V)$ we have $\|u\|_U \leq \|u + v\| + \|v\| \leq 2$, hence $f_{B(V)}(v^*) \leq \mu_U(2B(U)) = 2^a \cdot \mu_U(B(U))$ and

$$\mu_V(B(V)) \leq 2^a \cdot \mu_U(B(U)) \cdot \mu_{\alpha(V)}(B(\alpha(V))). \tag{6}$$

Furthermore, if $u_1, \ldots, u_a$ is a basis of $U \cap M$ and $u_1, \ldots, u_b$ is a basis of M, then $\alpha(u_{a+1}), \ldots, \alpha(u_b)$ is a basis of $\alpha(M)$ and

$$f_{V/M}(v^*) = \mu_U(\{u \in U : u + v \in V/M\}) = \mu_U(U/U \cap M)$$

for $v^* \in \alpha(V)/\alpha(M)$ and $f_{V/M}(v^*) = 0$ otherwise. Hence

$$\mu_V(V/M) = \int_{\alpha(V)} f_{V/M}(v^*)\, d\mu_{\alpha(V)}(v^*) = \mu_U(U/U \cap M) \cdot \mu_{\alpha(V)}(\alpha(V)/\alpha(M)). \tag{7}$$

Finally, by the definition of Vol(V), we have that $\mu_V(B(V))/\mu_V(V/M) = \mathrm{Vol}(V)$ and similarly for Vol(U) and Vol($\alpha(V)$). Combination of (6) and (7) now yields the statement of the lemma. □

Proof. (of Lemma 4) We have $\lambda_1(\alpha(V), M^*) \geq C^{-2}$, since for $0 \neq v^* \in \alpha(M) \subset \alpha(V)$ we have $v^* = \alpha(m)$ with $m \in M$ and

$$\|v^*\|^* = \inf_{u \in U} \|m + u\| \geq \frac{\|\alpha(m)\|_W}{\|\alpha\|} \geq C^{-2},$$

where $\|\ \|_W$ denotes the norm on W. By Minkowski's Theorem we obtain $\lambda_1(\alpha(V), M^*)^{b-a} \cdot \mathrm{Vol}(\alpha(V)) \leq 2^{b-a}$, so that $\mathrm{Vol}(\alpha(V)) \leq (2C^2)^{b-a}$. Further, $\lambda_b(V, M) \leq C$, and applying Minkowski's Theorem once more, we find that $\lambda_b(V, M)^b \cdot \mathrm{Vol}(V) \geq 2^b/b!$, whence $\mathrm{Vol}(V) \geq 2^b \cdot C^{-b}/b!$. Finally we have $\lambda_1(U, U \cap M)) \geq C^{-1}$ and, by Lemma 5, $\mathrm{Vol}(V) \leq 2^a \cdot \mathrm{Vol}(U) \cdot \mathrm{Vol}(\alpha(V))$, so that $\lambda_1(U, U \cap M)^i \cdot \lambda_{i+1}(U, U \cap M)^{a-i} \cdot \mathrm{Vol}(U) \leq 2^a$. Hence

$$\begin{aligned} \lambda_{i+1}(U, U \cap M) &\leq \left(2^a \cdot \mathrm{Vol}(U)^{-1} \cdot C^i\right)^{1/(a-i)} \leq \left(2^a \cdot \frac{\mathrm{Vol}(\alpha(V))}{\mathrm{Vol}(V)} \cdot C^i\right)^{1/(a-i)} \\ &\leq \left(2^a \cdot C^{i-2a+3b} \cdot b!\right)^{1/(a-i)} \leq \left(C^{3b} \cdot b!\right)^{1/(a-i)} \end{aligned}$$

for $0 \leq i \leq a - 1$. □

We give an example of how one obtains, in a general manner, a normed $\mathbb{R}$-vector space with a $\mathbb{Z}$-lattice.

8 Construction. Let K be a number field, $\mathcal{O}_K$ its ring of integers. Let X be a proper $\mathcal{O}_K$-scheme and let $\mathcal{L}$ be a metrized line bundle on X (i.e., $\mathcal{L}$ is an invertible $\mathcal{O}_X$-module equipped with a hermitean metric $\|\cdot\|_\sigma$ on $\mathcal{L}_\sigma$ for every embedding $\sigma: K \to \mathbb{C}$, see Chapter V, §4 for details). Then $\mathrm{H}^0(X,\mathcal{L})$ is a finitely generated $\mathcal{O}_K$-module by [27], Ch. III, Thm. 8.8 and Rem. 8.8.1; it is torsion free if X is flat over $\mathcal{O}_K$. We let $V = \mathrm{H}^0(X,\mathcal{L}) \otimes_{\mathbb{Z}} \mathbb{R}$. Then V is a $\mathbb{R}$-vector space and the image of the (canonical) map $\mathrm{H}^0(X,\mathcal{L}) \to \mathrm{H}^0(X,\mathcal{L}) \otimes_{\mathbb{Z}} \mathbb{R}$ is by construction a $\mathbb{Z}$-lattice in it. A norm on V can be obtained as follows. Since $K \otimes_{\mathbb{Q}} \mathbb{R} = \oplus_v K_v$ (the sum being taken over all infinite places). For v an infinite place of K let $V_v := \mathrm{H}^0(X_{K_v}, \mathcal{L}_{K_v})$, where X_{K_v} denotes the pullback of X to $\mathrm{Spec}(K_v)$ and $\mathcal{L}_{K_v}$ denotes the line bundle on X_{K_v} induced by $\mathcal{L}$. We have

$$V = \mathrm{H}^0(X,\mathcal{L}) \otimes_{\mathcal{O}_K} \mathcal{O}_K \otimes_{\mathbb{Z}} \mathbb{R} = \oplus_v \mathrm{H}^0(X,\mathcal{L}) \otimes_{\mathcal{O}_K} K_v = \oplus_v V_v$$

(for the last equality, use [27], Ch. III, Prop. 9.3). If $\sigma\colon K \to \mathbb{C}$ is a complex embedding giving the infinite place v, then on V_v we have the sup-norm over $X(\overline{K_v}) = X_\sigma(\mathbb{C})$. Note that by Chapter V, Def. 4.3, $\overline{\sigma}$ gives the same norm on V_v. On $V = \oplus_v V_v$ we take the max-norm associated to the norms on the V_v. In other words, this norm on V is just the sup-norm over $X(\mathbb{C}) = \coprod_\sigma X_\sigma(\mathbb{C})$. □

Chapter XI

Arithmetic Part of Faltings's Proof

by Bas Edixhoven

1 Introduction

In this chapter we will follow §5 of [22] quite closely. We start in the following situation:

> k is a number field, A_k is an abelian variety over k, $X_k \subset A_k$ is a subvariety such that $X_{\bar{k}}$ does not contain any translate of a positive dimensional abelian subvariety of $A_{\bar{k}}$, m is a sufficiently large integer as in Chapter IX, Lemma 1

Let R be the ring of integers in k. Since we want to apply Faltings's version of Siegel's Lemma (see Ch. X, Lemma 4) we need lattices in things like $\Gamma(X_k^m, \text{line bundle})$. We obtain such lattices as:

$$\Gamma(\text{proper model of } X_k^m \text{ over } R, \text{extension of line bundle}).$$

2 Construction of Proper Models

First we extend A_k to a group scheme over $\operatorname{Spec}(R)$. Among all such extensions there is a canonical "best" one, the so-called Néron model $\mathcal{A} \to \operatorname{Spec}(R)$ of A_k. It is characterized by the property that $\mathcal{A} \to \operatorname{Spec}(R)$ is smooth and that for every smooth morphism of schemes $T \to \operatorname{Spec}(R)$ the induced map $\mathcal{A}(T) \to A_k(T_k)$ is bijective. For more information and a proof of existence we refer to [11]; there is an "explicit" construction of $\mathcal{A}/R$ when A_k is a product of elliptic curves. We will use the following properties of $\mathcal{A} \to \operatorname{Spec}(R)$:

(i) it is a group scheme,

(ii) every $x \in A_k(k)$ extends uniquely to an $x \in \mathcal{A}(\operatorname{Spec}(R))$,

(iii) $\mathcal{A} \to \operatorname{Spec}(R)$ is quasi-projective ([11] §6.4, Thm. 1) and smooth.

We choose an embedding $i: \mathcal{A} \hookrightarrow \mathbb{P}_R^M$. Let $\overline{\mathcal{A}}$ be the closure of $\mathcal{A} \hookrightarrow \mathbb{P}_R^M \times_R \mathbb{P}_R^M$ by $P \mapsto (iP, i(-P))$, and let $A \to \overline{\mathcal{A}}$ be its normalization. Since $\overline{\mathcal{A}} \to \operatorname{Spec}(R)$ is projective and $A \to \overline{\mathcal{A}}$ is finite (by Nagata's theorem [47], Thm. 31.H), $A \to \operatorname{Spec}(R)$ is projective. By construction A contains $\mathcal{A}$ as an open subset, $A \times_{\operatorname{Spec}(R)} \operatorname{Spec}(k) = A_k$, and the automorphism $[-1]$ of A_k extends to A. Let $\mathcal{L}_0$ be a very ample line bundle on A. Then $\mathcal{L}_1 := \mathcal{L}_0 \otimes [-1]^*\mathcal{L}_0$ is a very ample line bundle on A whose restriction to A_k is symmetric. Finally let $\mathcal{L} = \mathcal{L}_1 \otimes p^*0^*\mathcal{L}_1^{-1}$, where $p\colon A \to \mathrm{Spec}(R)$. Then $\mathcal{L}$ is very ample on A and we have given isomorphisms $\mathcal{L} \xrightarrow{\sim} [-1]^*\mathcal{L}$ and $\mathcal{O}_{\mathrm{Spec}(R)} \xrightarrow{\sim} 0^*\mathcal{L}$.

For $1 \leq i,j \leq m$ we have the line bundles $\mathcal{L}_i := \mathrm{pr}_i^*\mathcal{L}$ and $\mathcal{P}_{i,j} := (\mathrm{pr}_i + \mathrm{pr}_j)^*\mathcal{L} \otimes \mathrm{pr}_i^*\mathcal{L}^{-1} \otimes \mathrm{pr}_j^*\mathcal{L}^{-1}$ on A_k^m. We choose positive rational numbers $\varepsilon \leq 1$ and s as in Chapter IX, Thm. 4: if $s_1, \ldots, s_m$ are positive rational numbers with $s_1/s_2 \geq s, \ldots, s_{m-1}/s_m \geq s$ and $\varepsilon' \leq \varepsilon$ then the restriction to X_k^m of the $\mathbb{Q}$-line bundle

$$\mathcal{L}(-\varepsilon', s_1, \ldots, s_m) := \mathcal{L}_1^{(1-\varepsilon')s_1^2} \otimes \bigotimes_{i=2}^{m-1} \mathcal{L}_i^{(2-\varepsilon')s_i^2} \otimes \mathcal{L}_m^{(1-\varepsilon')s_m^2} \otimes \bigotimes_{i=1}^{m-1} \mathcal{P}_{i,i+1}^{-s_i s_{i+1}}$$

on A_k^m is ample.

Let A^m denote the m-fold fibred product of $A \to \mathrm{Spec}(R)$ and let $\mathcal{L}_i := \mathrm{pr}_i^*\mathcal{L}$ on A^m. We extend the $\mathcal{P}_{i,j}$ to the open subset $\mathcal{A}^m$ of A^m by the same formula as above: $\mathcal{P}_{i,j} := (\mathrm{pr}_i + \mathrm{pr}_j)^*\mathcal{L} \otimes \mathrm{pr}_i^*\mathcal{L}^{-1} \otimes \mathrm{pr}_j^*\mathcal{L}^{-1}$. Using the construction described below it is easy to find a proper modification $B \to A^m$ such that 1) B is normal, 2) $B \to A^m$ is an isomorphism over $\mathcal{A}^m$ and 3) the $\mathcal{P}_{i,j}$ on $\mathcal{A}^m$ can be extended to line bundles on B.

2.1 Construction. Let X be an integral scheme, of finite type over $\mathbb{Z}$, $U \subset X$ open and normal, $D \subset U$ a closed subscheme whose sheaf of ideals I_D is an invertible $\mathcal{O}_U$-module. Let $\bar{D}$ be the scheme theoretic closure of D in X, let $\pi\colon \tilde{X} \to X$ be the blow up in $I_{\bar{D}}$, let X' be the normalization of the closure of $\pi^{-1}U$ in $\tilde{X}$ and let $\bar{D}'$ be the closure of $D' := \pi^{-1}D$ in X'. Then X' is normal, $X' \to X$ is proper and an isomorphism over U, and $I_{\bar{D}'}$ is an invertible $\mathcal{O}_{X'}$-module. □

We fix a choice of B and extensions of the $\mathcal{P}_{i,j}$ from $\mathcal{A}^m$ to B. The pullback of the $\mathcal{L}_i$ to B are still denoted $\mathcal{L}_i$. We let Y be the closure of X_k^m in B and we define $Y^o := Y \cap \mathcal{A}^m$. Note that every $x \in X_k^m(k)$ extends uniquely to an $x \in Y^o(R)$. We extend all line bundles $\mathcal{L}(-\varepsilon', s_1, \ldots, s_m)^d$ to B by their defining formula.

3 Applying Faltings's version of Siegel's Lemma

We fix norms on $\mathcal{L}$ at the infinite places. This gives norms on the $\mathcal{L}_i$, the $\mathcal{P}_{i,j}$ and the $\mathcal{L}(-\varepsilon', s_1, \ldots, s_m)^d$. In order to get some control over $\Gamma(Y, \mathcal{L}(-\varepsilon', s_1, \ldots, s_m)^d)$, using Faltings's version of Siegel's Lemma, we will embed $\mathcal{L}(-\varepsilon', s_1, \ldots, s_m)^d$ into $\oplus_1^a \otimes_{i=1}^m \mathcal{L}_i^{d_i}$ (a independent of the s_i, ε' and d) just as in the proof of Chapter IX, Thm. 4. However, this time we have to keep track of norms and denominators.

We fix a finite set of generators f_α, $\alpha \in I$, of the R-module $\Gamma(A, \mathcal{L})$. We choose an isomorphism ϕ on A_k^3:

$$(\mathrm{pr}_1 + \mathrm{pr}_2 + \mathrm{pr}_3)^*\mathcal{L} \xrightarrow{\sim} \bigotimes_{1 \leq i < j \leq 3} (\mathrm{pr}_i + \mathrm{pr}_j)^*\mathcal{L} \otimes \bigotimes_{1 \leq i \leq 3} \mathrm{pr}_i^*\mathcal{L}^{-1}$$

whose existence is guaranteed by the Theorem of the cube. Note that if ϕ' is another such isomorphism then $\phi' = u\phi$ for a unique $u \in k^*$. From ϕ we can construct isomorphisms $\phi_{a,b,i,j}$ on A_k^m for $a, b \in \mathbb{Z}$ and $1 \leq i,j \leq m$:

$$\phi_{a,b,i,j}\colon (a\mathrm{pr}_i + b\mathrm{pr}_j)^*\mathcal{L} \xrightarrow{\sim} \mathcal{L}_i^{a^2} \otimes \mathcal{L}_j^{b^2} \otimes \mathcal{P}_{i,j}^{ab}$$

as follows. Let $\mathcal{P} := (\mathrm{pr}_1 + \mathrm{pr}_2)^*\mathcal{L} \otimes \mathrm{pr}_1^*\mathcal{L}^{-1} \otimes \mathrm{pr}_2^*\mathcal{L}^{-1}$ on A_k^2. We have (on A_k^m and on A_k):

$$\phi\colon (\mathrm{pr}_i + \mathrm{pr}_j, \mathrm{pr}_l)^*\mathcal{P} \xrightarrow{\sim} (\mathrm{pr}_i, \mathrm{pr}_l)^*\mathcal{P} \otimes (\mathrm{pr}_j, \mathrm{pr}_l)^*\mathcal{P}$$

and

$$\phi^{-1}\colon [a+1]^*\mathcal{L} \xrightarrow{\sim} [a]^*\mathcal{L}^2 \otimes [a-1]^*\mathcal{L}^{-1} \otimes \mathcal{L}^2$$

And by definition: $(a\mathrm{pr}_i + b\mathrm{pr}_j)^*\mathcal{L} \xrightarrow{\sim} a\mathrm{pr}_i^*\mathcal{L} \otimes b\mathrm{pr}_j^*\mathcal{L} \otimes (a\mathrm{pr}_i, b\mathrm{pr}_j)^*\mathcal{P}$ on A_k^m.

3.1 Lemma. (Lemma 5.1 of [22]) *For any a, b, i, j let $\mathcal{F}_{a,b,i,j}$ be the subsheaf of $\mathcal{O}_B$-modules of the $\mathcal{O}_{A_k^m}$-module $\left(\mathcal{L}_i^{a^2} \otimes \mathcal{L}_j^{b^2} \otimes \mathcal{P}_{i,j}^{ab}\right) \otimes k$ generated by the $\phi_{a,b,i,j}(a\mathrm{pr}_i + b\mathrm{pr}_j)^* f_\alpha$, $\alpha \in I$. There exists $c_1 \in \mathbb{R}$ such that for all a, b, i, j there exists $r \in \mathbb{Z}$, $0 < r < \exp(c_1(1 + a^2 + b^2))$ with:*

1. $r \cdot \mathcal{F}_{a,b,i,j} \subset \mathcal{L}_i^{a^2} \otimes \mathcal{L}_j^{b^2} \otimes \mathcal{P}_{i,j}^{ab}$ *(on B)*
2. *on $\mathcal{A}^m$: $\mathcal{L}_i^{a^2} \otimes \mathcal{L}_j^{b^2} \otimes \mathcal{P}_{i,j}^{ab} \subset r^{-1} \cdot \mathcal{F}_{a,b,i,j}$*

Proof. Let us first check that the lemma holds for our choice of data if and only if it holds for any choice. If $\phi' = u^{-1}\phi$ with $u \in k^*$ then

$$\phi'_{a,b,i,j} = u^{a(a-1)/2 + b(b-1)/2 - a - b + 2}\phi_{a,b,i,j}$$

If $u\mathcal{L}_i \subset \mathcal{L}'_i \subset u^{-1}\mathcal{L}_i$ and $u\mathcal{P}_{i,j} \subset \mathcal{P}'_{i,j} \subset u^{-1}\mathcal{P}_{i,j}$ for all i, j then

$$u^{a^2+b^2+ab}\mathcal{L}_i^{a^2} \otimes \mathcal{L}_j^{b^2} \otimes \mathcal{P}_{i,j}^{ab} \subset \mathcal{L}_i'^{a^2} \otimes \mathcal{L}_j'^{b^2} \otimes \mathcal{P}_{i,j}'^{ab} \subset u^{-(a^2+b^2+ab)}\mathcal{L}_i^{a^2} \otimes \mathcal{L}_j^{b^2} \otimes \mathcal{P}_{i,j}^{ab}$$

If $\pi\colon B' \to B$ is another model satisfying the same conditions as B, then for any line bundle $\mathcal{M}$ on B we have: $\Gamma(B', \pi^*\mathcal{M}) = \Gamma(B, \pi_*\pi^*\mathcal{M}) = \Gamma(B, \mathcal{M})$; this shows that c_1 works for B' if and only if c_1 works for B (if we take $\mathcal{P}'_{i,j} = \pi^*\mathcal{P}_{i,j}$).

Let us now prove statements (1) and (2) on $\mathcal{A}^m$. By [50] Lemme II, 1.2.1, there exists $d_0 > 0$ such that $\mathcal{L}^{d_0}|_{A_k}$ can be extended to a $\mathcal{L}'$ on $\mathcal{A}$ for which some isomorphism ϕ' on A_k^3 extends to an isomorphism over $\mathcal{A}^3$. This means that the lemma is true for $\mathcal{L}'$ on $\mathcal{A}^m$, hence for $\mathcal{L}^{d_0}$ and hence for $\mathcal{L}$ itself (on $\mathcal{A}^m$).

To finish the proof we have to consider pole orders of the $\phi_{a,b,i,j}(a\mathrm{pr}_i + b\mathrm{pr}_j)^* f_\alpha$ along the (finitely many) divisors of B contained in $B - \mathcal{A}^m$; these divisors are irreducible components of closed fibres of $B \to \mathrm{Spec}(R)$. Let V be the local ring on B at the generic point of such a divisor. Then V is a discrete valuation ring and pr_i and pr_j define V-valued points of A. The lemma is then proved by base changing from R to V, replacing A_V by the Néron model of A over V, and applying the arguments above. Note that those arguments give upper bounds on pole and zero orders on the whole model; that upper bounds on the pole orders on the whole model give upper bounds on the pole orders on V-valued points but that the same is not true for the order of zeros since the divisors of zeros of the $\phi_{a,b,i,j}(a\mathrm{pr}_i + b\mathrm{pr}_j)^* f_\alpha$ are not necessarily supported on closed fibres. □

Let $s_1, \ldots, s_m$ be positive rational numbers, let $0 \leq \varepsilon' \leq \varepsilon$ and let $d \in \mathbb{Z}$, $d > 0$, be a sufficiently divisible square. Let $d_i := 4ds_i^2$. Let α_i $(1 \leq i \leq m-1)$ and β_i $(1 \leq i \leq m)$ be arbitrary elements of the set I of labels of the f_α. Multiplication by the product of the $r_i \cdot \phi_{s_i\sqrt{d}, s_{i+1}\sqrt{d}, i, i+1}(s_i\sqrt{d}\mathrm{pr}_i + s_{i+1}\sqrt{d}\mathrm{pr}_{i+1})^* f_{\alpha_i}$ (with the r_i as given by Lemma 3.1), the $\mathrm{pr}_i^* f_{\beta_i}^{\varepsilon' ds_i^2}$, $\mathrm{pr}_1^* f_{\beta_1}^{2ds_1^2}$ and $\mathrm{pr}_m^* f_{\beta_m}^{2ds_m^2}$ is a map from $\mathcal{L}(-\varepsilon', s_1, \ldots, s_m)^d$

to $\bigotimes_{i=1}^m \mathcal{L}_i^{d_i}$. Since the f_α ($\alpha \in I$) are generators of $\Gamma(A, \mathcal{L})$, all such maps together give an embedding on B:

$$\rho: \mathcal{L}(-\varepsilon', s_1, \ldots, s_m)^d \hookrightarrow \bigoplus_{j=1}^{a} \bigotimes_{i=1}^{m} \mathcal{L}_i^{d_i}$$

where $a = \#I^{2m-1}$. By construction, this embedding remains an embedding after restriction to Y; however, only on $B_k = A_k^m$ we know that it is an embedding without zeros (that is, locally split).

3.2 Constructions. 1. Let A be a commutative ring, $a_1, \ldots, a_n$, $r \in A$ such that $r \in \sum_i Aa_i$. Then the homology of the (start of the Koszul) complex:

$$0 \to A \to A^n \to A^{n^2}, \qquad x \mapsto (a_i x)_i, \qquad (x_i)_i \mapsto (a_j x_i - a_i x_j)_{i,j}$$

is annihilated by r. Proof: choose b_i such that $\sum a_i b_i = r$. Then use the "homotopy": $(x_i)_i \mapsto \sum_i b_i x_i$, $(x_{i,j})_{i,j} \mapsto (\sum_j b_j x_{i,j})_i$.

2. Let X be a scheme, $\mathcal{L}$ and $\mathcal{M}$ invertible $\mathcal{O}_X$-modules, $r \in \mathcal{O}_X(X)$, $\rho_j: \mathcal{L} \to \mathcal{M}$ ($1 \leq j \leq a$) a finite set of morphisms such that $r\mathcal{M} \subset \sum_j \rho_j(\mathcal{L})$. Twisting by $\mathcal{L}^{-1}$ gives $\rho_j: \mathcal{O}_X \to \mathcal{M} \otimes \mathcal{L}^{-1}$. As in (1) we get a complex $0 \to \mathcal{O}_X \to \oplus_{j=1}^a (\mathcal{M} \otimes \mathcal{L}^{-1}) \to \oplus_{j=1}^{a^2} (\mathcal{M} \otimes \mathcal{L}^{-1})^2$. Twisting by $\mathcal{L}$ gives:

$$0 \to \mathcal{L} \to \oplus_{j=1}^a \mathcal{M} \to \oplus_{j=1}^{a^2} \mathcal{M}^2 \otimes \mathcal{L}^{-1}$$

The homology of this complex is annihilated by r. □

We apply these constructions to the embeddings $\rho_j: \mathcal{L}(-\varepsilon', s_1, \ldots, s_m)^d \to \otimes_{i=1}^m \mathcal{L}_i^{d_i}$. In this case we can even embed $\mathcal{L}(-\varepsilon', s_1, \ldots, s_m)^{-d}$ into $\oplus_{j=1}^a \otimes_{i=1}^m \mathcal{L}_i^{d_i}$ by multiplication by the $r_i \cdot \phi_{s_i\sqrt{d}, -s_{i+1}\sqrt{d}, i, i+1}(s_i\sqrt{d}\mathrm{pr}_i - s_{i+1}\sqrt{d}\mathrm{pr}_{i+1})^* f_{\alpha_i}$ and by the $\mathrm{pr}_i^* f_{\beta_i}^{(4-\varepsilon')ds_i^2}$. All together we get a complex:

$$0 \to \mathcal{L}(-\varepsilon', s_1, \ldots, s_m)^d \to \bigoplus_{j=1}^{a} \bigotimes_{i=1}^{m} \mathcal{L}_i^{d_i} \to \bigoplus_{j=1}^{a^3} \bigotimes_{i=1}^{m} \mathcal{L}_i^{3d_i}$$

of sheaves on Y whose homology on Y^o is annihilated by some $r \in \mathbb{Z}$ with $0 < r < \exp(c_1 \sum_{i=1}^m d_i)$.

3.3 Proposition. (Proposition 5.2 of [22]) *There exists $c_2 \in \mathbb{R}$ such that for all $s_1, \ldots, s_m$, ε' positive rational numbers with $0 < \varepsilon' \leq \varepsilon$, $d \in \mathbb{Z}$ square and sufficiently divisible there exists an exact sequence:*

$$0 \to \Gamma(X_k^m, \mathcal{L}(-\varepsilon', s_1, \ldots, s_m)^d) \to \Gamma(X_k^m, \bigoplus_{j=1}^{a} \bigotimes_{i=1}^{m} \mathcal{L}_i^{d_i}) \to \Gamma(X_k^m, \bigoplus_{j=1}^{a^3} \bigotimes_{i=1}^{m} \mathcal{L}_i^{3d_i})$$

such that:

1. *the norms of the maps in the exact sequence and the difference between the norm on $\Gamma(X_k^m, \mathcal{L}(-\varepsilon', s_1, \ldots, s_m)^d)$ and that induced on it from $\Gamma(X_k^m, \oplus_{j=1}^a \otimes_{i=1}^m \mathcal{L}_i^{d_i})$ are bounded by $\exp(c_2 \sum_i d_i)$,*

2. *if a section f of $\mathcal{L}(-\varepsilon', s_1, \ldots, s_m)^d$ over X_k^m maps to $\Gamma(Y^o, \oplus_\alpha \otimes_{i=1}^m \mathcal{L}_i^{d_i})$ then there exists $r \in \mathbb{Z}$ with $0 < r < \exp(c_2 \sum_i d_i)$ such that $rf \in \Gamma(Y^o, \mathcal{L}(-\varepsilon', s_1, \ldots, s_m)^d)$.*

Proof. The exact sequence is obtained by applying $\Gamma(X_k^m, -)$ to the complex above. It remains to prove the statements concerning norms. To make sense of these statements we must say which norms we are talking about. The norms on the line bundles are obtained from those on $\mathcal{L}$. On k_v-vector spaces such as

$$\Gamma(X_k^m, \text{metrized vector bundle}) \otimes_k k_v$$

we take the supremum norm over the complex valued points. On finite direct sums of normed k_v-vector spaces we take the maximum norm. Finally, if V is a k-vector space with given norms on the $V \otimes_k k_v$ then we take the maximum norm on

$$V \otimes_{\mathbb{Q}} \mathbb{R} = V \otimes_k k \otimes_{\mathbb{Q}} \mathbb{R} = V \otimes_k (\oplus_v k_v) = \oplus_v (V \otimes_k k_v)$$

Note that the norm we have put on $(\oplus_j \Gamma(X_k^m, \mathcal{F}_j)) \otimes_{\mathbb{Q}} \mathbb{R}$ can be viewed as the supremum norm on $\Gamma(X_k^m \times_{\mathrm{Spec}(\mathbb{Q})} \mathrm{Spec}(\mathbb{R}), \oplus_j \mathcal{F}_j)$, the supremum being taken over the $P \in X_k^m(\mathbb{C})$. The proof concerning the statements about norms is analogous to the arguments to bound poles and zeros at the finite places. The analog of Lemma 3.1 is independent of the choice of metrics on $\mathcal{L}$, and for a particular choice (metrics with translation invariant curvature forms) ϕ is an isometry. □

We want to apply the following lemma (for a proof see Ch. X, Lemma 4).

3.4 Lemma. (Prop. 2.18 of [22]) *Let $\rho: V \to W$ be a linear map between normed finite dimensional $\mathbb{R}$-vector spaces with lattices M and N such that $\rho(M) \subset N$. Let $U := \ker(\rho)$ and $L := M \cap U$; then $L \subset U$ is a lattice. Assume that for some $C \geq 2$ we have: 1. ρ has norm $\leq C$, 2. M is generated by elements of norm $\leq C$, 3. any non-trivial element of M or of N has norm $\geq C^{-1}$. Then for any proper subspace $U_0 \subset U$ there exists an element $f \in L$, $f \notin U_0$ such that $\|f\| \leq (C^3 \dim V)^{\dim(V)/\mathrm{codim}(U_0)}$.* □

Now let $U := \Gamma(X_k^m, \mathcal{L}(-\varepsilon', s_1, \ldots, s_m)^d) \otimes_{\mathbb{Q}} \mathbb{R}$, $V := \Gamma(X_k^m, \oplus_{j=1}^a \otimes_{i=1}^m \mathcal{L}_i^{d_i}) \otimes_{\mathbb{Q}} \mathbb{R}$ and $W := \Gamma(X_k^m, \oplus_{j=1}^{a^3} \otimes_{i=1}^m \mathcal{L}_i^{3d_i}) \otimes_{\mathbb{Q}} \mathbb{R}$. The lattices M and N are $\Gamma(Y, \oplus_{j=1}^a \otimes_{i=1}^m \mathcal{L}_i^{d_i})$ and $\Gamma(Y, \oplus_{j=1}^{a^3} \otimes_{i=1}^m \mathcal{L}_i^{3d_i})$. Condition 1 is satisfied with $C = \exp(c_2 \sum_i d_i)$. Condition 2 is satisfied with C of the form $\exp(c_3 \sum_i d_i)$ (with c_3 independent of the s_i) because $S := \oplus_{d_1,\ldots,d_m \geq 0} \Gamma(Y, \otimes_{i=1}^m \mathcal{L}_i^{d_i})$ is a finitely generated R-algebra by the following argument. The morphism $f: Y \to X^m$ is proper and each $\mathcal{L}_i$ is of the form $f^*\mathcal{L}_i'$, hence $S = \oplus_{d_1,\ldots,d_m \geq 0} \Gamma(X^m, (f_*\mathcal{O}_Y) \otimes \otimes_{i=1}^m \mathcal{L}_i'^{d_i})$ by the projection formula ([27], Ch. 2, Exc. 5.1). Let $\iota: X^m \hookrightarrow P := (\mathbb{P}_R^M)^m$ denote a closed immersion induced by $\mathcal{L}$ on each factor. Then $S = \oplus_{d_1,\ldots,d_m \geq 0} \Gamma(P, \mathcal{F}(d_1, \ldots, d_m))$, with $\mathcal{F}$ the coherent $\mathcal{O}_P$-module $\iota_* f_* \mathcal{O}_Y$. One then shows that S is a finitely generated module over the multi-homogeneous coordinate ring of P.

The proof of condition 3 that we give is copied from [81], Lemma 13.9.

3.5 Lemma. *There exists $c_4 \in \mathbb{R}$ such that for all $d_i \geq 0$ and for all non-zero $f \in \Gamma(Y, \otimes_{i=1}^m \mathcal{L}_i^{d_i})$ we have $\|f\| \geq \exp(-c_4 \sum_i d_i)$.*

Proof. Let $[k' : k] < \infty$, let R' be the ring of integers in k', $P = (P_1, \ldots, P_m) \in X_k^m(k') = Y(R')$ such that $P^*(f) \neq 0$. Then:

$$h_{\otimes \mathcal{L}_i^{d_i}}(P) = [k' : \mathbb{Q}]^{-1} \deg_{R'}(P^* \otimes_i \mathcal{L}_i^{d_i}) = \sum_{i=1}^m d_i h_{\mathcal{L}}(P_i)$$

On the other hand:

$$h_{\otimes \mathcal{L}_i^{d_i}}(P) = [k':\mathbb{Q}]^{-1}\left(\log \#\left(\left(P^* \otimes_i \mathcal{L}_i^{d_i}\right)/R' \cdot P^*(f)\right) - \sum_{v'|\infty} \epsilon(v') \log \|f(P)\|_{v'}\right)$$

where $\epsilon(v') = 1$ if v' is real and $\epsilon(v') = 2$ if v' is complex. Hence:

$$[k':\mathbb{Q}]^{-1} \sum_{v'|\infty} \epsilon(v') \log \|f(P)\|_{v'} \geq -\sum_{i=1}^{m} d_i h_{\mathcal{L}}(P_i)$$

Let $\pi_i \colon X_k \to \mathbb{P}^{\dim X_k}$ be associated to $\mathcal{L}$: $\pi_i^*\mathcal{O}(1) \cong \mathcal{L}$, and let h be the naive height function on $\mathbb{P}^{\dim X_k}(\bar{k})$. Then there exists c_4 such that $|h_{\mathcal{L}}(P_i) - h(\pi_i(P_i))| \leq c_4$ for all k' and all P_i. Now we take the P_i such that all coordinates of the $\pi_i(P_i)$ are roots of unity; then $h(\pi_i P_i) = 0$. Note that these points are Zariski dense. So we have:

$$[k':\mathbb{Q}]^{-1} \sum_{v'|\infty} \epsilon(v') \log \|f(P)\|_{v'} \geq -c_4 \sum_{i=1}^{m} d_i$$

It follows that there exists an infinite place v of k with $\|f\|_{v,\sup} \geq \exp(-c_4 \sum_i d_i)$. □

To get a suitable subspace U_0 of U we put $\varepsilon' = \varepsilon - \sigma$ with $0 < \sigma < \varepsilon$, let $x = (x_1, \ldots, x_m) \in X_k^m(k)$ be a smooth point and let

$$U_0 := \{f \in \Gamma(X_k^m, \mathcal{L}(\sigma - \varepsilon, s_1, \ldots, s_m)^d) \mid \operatorname{index}(x, f) \geq \sigma\}$$

For this to be useful we need a lower bound for $\operatorname{codim}(U_0)$. Note that:

$$\mathcal{L}(\sigma - \varepsilon, s_1, \ldots, s_m)^d = \mathcal{L}(-\varepsilon, s_1, \ldots, s_m)^d \otimes \bigotimes_{i=1}^{m} \mathcal{L}_i^{d\sigma s_i^2}$$

Because $\mathcal{L}(-\varepsilon, s_1, \ldots, s_m)$ is ample on X_k^m, $\mathcal{L}(-\varepsilon, s_1, \ldots, s_m)^d$ has a global section that does not vanish at x. Because $\mathcal{L}$ is very ample on X_k there are global sections of $\mathcal{L}^{d\sigma s_i^2}$ on X_k with prescribed Taylor expansion at x_i up to order $d\sigma s_i^2 = d_i\sigma/4$. Namely: "$\mathcal{L}$ very ample on X_k" is equivalent to "$\Gamma(X_k, \mathcal{L})$ separates points and tangent vectors" which implies "$\Gamma(X_k, \mathcal{L}) \to \mathcal{O}_{X_k, x_i}/m_{x_i}^2$ is surjective for all i" which implies "$\Gamma(X_k, \mathcal{L}^{\sigma d_i/4}) \to \mathcal{O}_{X_k, x_i}/m_{x_i}^{1+\sigma d_i/4}$ is surjective". It follows that $\operatorname{codim} U_0 \geq c_5(\sigma^m \prod_i d_i)^{\dim X_k}$, with $c_5 > 0$ depending only on m and $\dim X_k$. In Chapter VII it was proved that $\dim V \leq c_6(\prod_i d_i)^{\dim X_k}$ with c_6 independent of the d_i (and of σ of course). It follows that $(\dim V)/(\operatorname{codim} U_0) \leq c_7 \sigma^{-m \dim X_k}$ with c_7 independent of the d_i and of σ. Prop. 3.3 and Lemma 3.4 give us $f \neq 0$ in $\Gamma(Y, \mathcal{L}(\sigma - \varepsilon, s_1, \ldots, s_m)^d)$ with index $< \sigma$ at x and $\log \|f\| \leq c_8 \sigma^{-m \dim X_k} \sum_{i=1}^m d_i$ (where c_8 is independent of the d_i and of σ). We have proved:

3.6 Theorem. (Theorem 5.3 of [22]) *Let σ be a rational number with $0 < \sigma < \varepsilon$. There exists a real number $c(\sigma)$, depending only on σ (recall that ε is fixed), with the following property. For any smooth k-rational point x in X_k^m, any sequence $s_1, \ldots, s_m$ of positive rational numbers with $s_1/s_2 \geq s, \ldots, s_{m-1}/s_m \geq s$, and any square integer d which is big enough and sufficiently divisible there exists $f \in \Gamma\left(Y^\circ, \mathcal{L}(\sigma - \varepsilon, s_1, \ldots, s_m)^d\right)$ with index less than σ at x and norm at the infinite places bounded by $\exp(c(\sigma) \sum_{i=1}^m d_i)$, where $d_i = 4ds_i^2$.* □

4 Leading Terms and Differential Operators

This section more or less follows part of a manuscript [33] on [22] written by L. Lafforgue in 1991 (not published), see also [34] and [35]. We will need some properties of leading terms in Taylor expansions in the following situation. Let X be an integral separated R-scheme and $x\colon \mathrm{Spec}(R) \to X$ an R-valued point of X. Let $\partial_1, \ldots, \partial_m$ be R-derivations of $\mathcal{O}_X$, let $\mathcal{L}$ be an invertible $\mathcal{O}_X$-module and f a section of $\mathcal{L}$ vanishing up to order e along x, that is, $f \in \Gamma(X, I^e\mathcal{L})$, where $I \subset \mathcal{O}_X$ is the ideal sheaf of x. Then for integers $e_1, \ldots, e_m \geq 0$ with $\sum_i e_i = e$ we can define $(\partial_1^{e_1} \cdots \partial_m^{e_m} f)(x)$ in $\Gamma(\mathrm{Spec}(R), x^*\mathcal{L})$ as follows: locally on X write $f = gf_1$ with f_1 a generator of $\mathcal{L}$ and put:

$$(\partial_1^{e_1} \cdots \partial_m^{e_m} f)(x) := (\partial_1^{e_1} \cdots \partial_m^{e_m} g)(x) f_1(x)$$

(one easily checks that the right hand side does not depend on the choice of f_1, so the local construction "glues"). Also, it is important to note that $(\partial_1^{e_1} \cdots \partial_m^{e_m} f)(x)$ does not depend on the order in which the derivations are applied, so that we can write $(\prod_i \partial_i^{e_i} \cdot f)(x)$ for it, and that for derivations $\partial'_1, \ldots, \partial'_m$ locally of the form $\partial'_i = \partial_i +$ terms like $g \cdot \partial$ with $g \in \Gamma(X, I)$ and $\partial\colon \mathcal{O}_X \to \mathcal{O}_X$ an R-derivation, one has $(\prod_i \partial_i'^{e_i} \cdot f)(x) = (\prod_i \partial_i^{e_i} \cdot f)(x)$.

4.1 Lemma. *In this situation,* $(\prod_i (\partial_i^{e_i}/e_i!) \cdot f)(x)$, *which is a priori only a section of* $x^*\mathcal{L}$ *over* $\mathrm{Spec}(k)$, *is actually a section of* $x^*\mathcal{L}$ *over* $\mathrm{Spec}(R)$.

Proof. It suffices to verify this locally, so we may assume that X is affine, that $\mathcal{L}$ is trivialised by f_1 and that $f = gf_1$. Since f vanishes up to order e at x, g can be written as a sum of products $g_1 \cdots g_e$ with the g_i in $\Gamma(X, I)$. It suffices to verify the claim for each term, so we may assume that g is one such term, say $g = g_1 \cdots g_e$. Now consider $((\prod_{i=1}^m \partial_i^{e_i}) g_1 \cdots g_e)(x)$. This expression can be expanded by applying the product rule $(\partial(fg) = f\partial(g) + g\partial(f))$ as many times as possible. Since we evaluate at x, and the $g_i(x)$ are zero, only terms in which all g_i have been derived can give a non-zero contribution. As $\sum_i e_i = e$, we see that $((\prod_{i=1}^m \partial_i^{e_i}) g_1 \cdots g_e)(x)$ equals the sum over all partitions of $\{1, 2, \ldots, e\}$ into sets $S_1, \ldots, S_m$ of cardinalities $e_1, \ldots, e_m$, of the expresion

$$\prod_{i=1}^m \left(\partial_i^{e_i} \big(\prod_{j \in S_i} g_j \big) \right)(x) = \prod_{i=1}^m \left(e_i! \prod_{j \in S_i} (\partial_i g_j)(x) \right)$$

so the claim that $(\prod_i (\partial_i^{e_i}/e_i!) \cdot f)(x)$ is integral has now been proved. □

However, in §5 we cannot apply this result directly, because we will only know that f vanishes at x up to order e on the generic fibre X_k of X. The problem is then that in general the scheme theoretic closure in X of the closed subscheme of X_k defined by $(I \cdot \mathcal{O}_{X_k})^e$ is strictly contained in the closed subscheme of X defined by I^e (in other words: $I^e \neq \mathcal{O}_X \cap (I \cdot \mathcal{O}_{X_k})^e$). Yet another complication will be that we will work in a product situation with weighted degrees for differential operators. The result we need is the following.

Let $m \geq 1$. For $1 \leq i \leq m$ let $p_i\colon X_i \to P_i$ be a morphism of integral separated R-schemes of finite type, $\mathcal{M}_i$ an invertible $\mathcal{O}_{X_i}$-module, $G_i \in \Gamma(X_i, \mathcal{M}_i)$ a global section annihilating $\Omega^1_{X_i/P_i}$ and $\partial_{i,j}$, $1 \leq j \leq n_i$, some set of R-derivations on P_i. Let

$X := X_1 \times \cdots \times X_m$, $P := P_1 \times \cdots \times P_m$ (fibred products over Spec(R)), and $p: X \to P$ the product of the p_i. Let Y be a separated R-scheme with a R-morphism $q: Y \to X$ that is an isomorphism on the generic fibres: $q_k: Y_k \tilde{\to} X_k$. Let $\mathcal{N}$ be an invertible $\mathcal{O}_Y$-module and $N \in \Gamma(Y, \mathcal{N})$ a global section annihilating $\Omega^1_{Y/X}$. Let $x \in Y(R)$ and write $q(x) = (x_1, \ldots, x_m)$ with $x_i \in X_i(R)$. Suppose that for all i the morphism of k-schemes $p_k: X_{i,k} \to P_{i,k}$ is etale at $x_{i,k}$, and that $P_i \to \mathrm{Spec}(R)$ is smooth along $p_i(x_i)$. Let $\mathcal{L}$ be an invertible $\mathcal{O}_Y$-module, $f \in \Gamma(Y, \mathcal{L})$ and let σ be the index of $f|_{Y_k}$ at x_k with respect to some weights $d_1, \ldots, d_m > 0$. Finally let $e_{i,j} \geq 0$ be given with $\sum_{i,j} e_{i,j}/d_i = \sigma$. Since the $p_{i,k}: X_{i,k} \to P_{i,k}$ are etale at $x_{i,k}$, the $\partial_{i,j}$ can be uniquely lifted to derivations $\partial_{i,j}$ on a neighborhood of $x_{i,k}$ in $X_{i,k}$. Since $Y_k = X_k = \prod_i X_{i,k}$ we can view the $\partial_{i,j}$ as derivations in the ith direction on a neighborhood of x_k in Y_k. As usual, one can define $(\prod_{i,j} \partial_{i,j}^{e_{i,j}} f)(x_k)$ in $x_k^*\mathcal{L}$ to be $(\prod_{i,j} \partial_{i,j}^{e_{i,j}} g)(x_k) \cdot f_1(x_k)$, where f_1 is a generator of $\mathcal{L}_{x_k}$ and $f = g f_1$.

4.2 Lemma. *In this situation,*

$$\prod_{i,j} ((NG_i)^{e_{i,j}}/e_{i,j}!) \cdot \left(\left(\prod_{i,j} \partial_{i,j}^{e_{i,j}} \right) f \right) (x_k)$$

extends to a global section over Spec(R) *of* $x^*\mathcal{L} \otimes x^*\mathcal{N}^e \otimes \bigotimes_{i=1}^m x_i^*\mathcal{M}_i^{e_i}$, *where* $e_i = \sum_j e_{i,j}$ *and* $e = \sum_i e_i$.

Proof. It suffices to check the statement locally, so after localizing R we may assume that $\mathcal{L}$ has generator f_1, that $f = g f_1$ and that the $\mathcal{M}_i$ and $\mathcal{N}$ are trivial. Let $J_i \subset \mathcal{O}_{P_i}$ be the ideal sheaf of $p_i(x_i)$ and let $J \subset \mathcal{O}_P$ be the ideal sheaf of $p(q(x))$. Let $M_i \subset \mathcal{O}_{X_i}$ be the ideal of x_i and $M \subset \mathcal{O}_X$ the ideal of $q(x)$. Let $I \subset \mathcal{O}_Y$ be the ideal of x and $I_i := \mathcal{O}_Y \cdot q^\# M_i$. The closed subscheme of Y defined by the ideal $I^{e+1} + \sum_{i=1}^m I_i^{e_i+1}$ is finite over Spec(R), hence affine, so isomorphic to Spec(A) for some R-algebra A. The ideals of A induced by I and the I_i will be denoted by the same symbols. For each i, let Spec(B_i) denote the closed subscheme of X_i defined by $M_i^{e_i+1}$. Let $B := \bigotimes_{i=1}^m B_i$ (tensor product over R); then Spec(B) is the closed subscheme of X defined by $\sum_i M_i^{e_i+1}$. Finally, let Spec(C_i) be the closed subscheme of P_i defined by $J_i^{e_i+1}$ and let $C := \bigotimes_{i=1}^m C_i$ (tensor product over R). Then we have a commutative diagram

$$\begin{array}{ccccc} Y & \to & X & \to & P \\ \uparrow & & \uparrow & & \uparrow \\ \mathrm{Spec}(A) & \to & \mathrm{Spec}(B) & \to & \mathrm{Spec}(C) \end{array}$$

The rings A, B and C are R-modules of finite type, the morphisms $C \to B \to A$ induce isomorphisms $C \otimes_R k \tilde{\to} B \otimes_R k \tilde{\to} A \otimes_R k$. Because of the smoothness of the $P_i \to \mathrm{Spec}(R)$ at the $p(x_i)$, the rings C_i and C are torsion free R-modules.

4.3 Lemma. *Let* $\overline{B}$ *be the image of* B *in* A, *then* $A/\overline{B}$ *is annihilated by* N^e.

Proof. We know that N annihilates $\Omega^1_{Y/X}$, hence also $x^*\Omega^1_{Y/X} = I/(I^2 + A\overline{M})$, where $\overline{M}$ is the image of M in A; note that $\overline{M} \subset I$. Since $A = R + I$, we have $I^2 + A \cdot \overline{M} = I^2 + \overline{M} + I\overline{M} = I^2 + \overline{M}$. It follows that $NI \subset I^2 + \overline{M}$. Induction shows $N^e I \subset I^{e+1} + \overline{M} = \overline{M}$, hence $N^e A = N^e(R + I) \subset R + \overline{M} = \overline{B}$. □

4.4 Lemma. *For all i, the morphism $C_i \to B_i$ is injective, and B_i/C_i is annihilated by $G_i^{e_i}$.*

Proof. The morphism $C_i \to B_i$ is injective since C_i is torsion free and $C_{i,k} \tilde{\to} B_{i,k}$. The proof that $G_i^{e_i}$ annihilates B_i/C_i is the same as the proof of the previous lemma. □

4.5 Lemma. *The morphism $C \to B$ is injective, and B/C is annihilated by $\prod_{i=1}^m G_i^{e_i}$.*

Proof. The morphism $C \to B$ is injective since C is torsion free and $C_k \tilde{\to} B_k$. Let us consider the filtration

$$B = B^1 \supset B^2 \supset \cdots \supset B^m = C$$

where B^j is the image in $B = B_1 \otimes \cdots \otimes B_m$ of $\bigotimes_{i<j} C_i \otimes \bigotimes_{i \geq j} B_i$. The successive quotient B^j/B^{j+1} of this filtration is a quotient of $\bigotimes_{i<j} C_i \otimes (B_j/C_j) \otimes \bigotimes_{i>j} B_j$, hence by the previous lemma it is annihilated by $G_j^{e_j}$. The lemma follows. □

We have now proved that $h := (N^e \prod_{i=1}^m G_i^{e_i})g$, modulo $I^{e+1} + \sum_{i=1}^m I_i^{e_i+1}$, is in C. Note that $(\prod_{i,j} \partial_{i,j}^{e_{i,j}} g')(x)$ is zero for all $g' \in I^{e+1} + \sum_{i=1}^m I_i^{e_i+1}$. Let us consider h as an element of C. Since the index at x_k of $f|_{Y_k}$ is σ, it follows that the image of h in C_k is in the ideal $C_k \cdot J_1^{e_1} \cdot \cdots \cdot J_m^{e_m}$. Since the ring $C/(C \cdot J_1^{e_1} \cdot \cdots \cdot J_m^{e_m})$ is torsion free (this follows again from the smoothness of $P \to \mathrm{Spec}(R)$ at $p(q(x))$), h is actually an element of $C \cdot J_1^{e_1} \cdot \cdots \cdot J_m^{e_m}$. The method of the proof of Lemma 4.1, plus the fact that the $\partial_{i,j}$ are derivations on P in the direction of P_i, show that $((\prod_{i,j} \partial_{i,j}^{e_{i,j}}/e_{i,j}!)h)(p(q(x)))$ is integral. □

5 Proof of the Main Theorem

5.1 Theorem. (Theorem I of [22]) *Let k be a number field, A_k an abelian variety over k and $X_k \subset A_k$ a closed subvariety such that $X_{\bar{k}}$ does not contain any translate of a positive dimensional abelian subvariety of $A_{\bar{k}}$. Then $X_k(k)$ is finite.*

Proof. The proof is by contradiction: we suppose that $X_k(k)$ is *not* finite. We may also suppose that $X_k(k)$ is Zariski-dense in X_k, because we can replace X_k by a positive dimensional irreducible component of the Zariski-closure of $X_k(k)$.

The situation is now as in §§1–3: A, $\mathcal{L}$, m, B, Y, the $\mathcal{P}_{i,j}$, ε and s are fixed. We embed X_k into some projective space over k by choosing a basis of $\Gamma(X_k, \mathcal{L}|_{X_k})$. Composing this embedding with a suitable projection as in Chapter IX, Lemma 5 we get a finite morphism $\mathrm{pr}: X_k \to \mathbb{P}_k^n$, where $n = \dim X_k$. Note that $\mathcal{L} = \mathrm{pr}^*\mathcal{O}(1)$. Let X denote the closure of X_k in A and let $\widetilde{X} \to X$ be a proper modification such that $\widetilde{X}_k = X_k$ and that $\mathrm{pr}: X_k \to \mathbb{P}_k^n$ extends to $\mathrm{pr}: \widetilde{X} \to \mathbb{P}_R^n$ (e.g., one can take $\widetilde{X}$ to be the closure in $X \times_{\mathrm{Spec}(R)} \mathbb{P}_R^n$ of the graph of $\mathrm{pr}: X_k \to \mathbb{P}_k^n$). We choose an integer g and a non-zero G in $\Gamma(\mathbb{P}_R^n, \mathcal{O}(g))$ which annihilates $\Omega^1_{\widetilde{X}/\mathbb{P}_R^n}$ (Lemma 5 of Chapter IX shows the existence of such a G annihilating $\Omega^1_{X_k/\mathbb{P}_k^n}$; multiplying such a G by a suitable integer gives the G we want). Let $\widetilde{Y} \to Y$ be a proper modification such that $\widetilde{Y}_k = Y_k$ and that the projections $\mathrm{pr}_i: Y_k = X_k^m \to X_k$ on the ith factor extend to $\mathrm{pr}_i: \widetilde{Y} \to \widetilde{X}$. In particular, $\widetilde{Y}$ is a model of X_k^m dominating both Y and $\widetilde{X}^m$ ($m$fold fibred product over Spec(R)). Since $\widetilde{Y} \to \widetilde{X}^m$ is an isomorphism on the generic fibres, we can choose an integer $N > 0$ such that N annihilates $\Omega^1_{\widetilde{Y}/\widetilde{X}^m}$. For $1 \leq i \leq m$ let $P_i := \mathbb{P}^n_k$ and $P_{i,R} := \mathbb{P}^n_R$. Let $p_i: \widetilde{Y} \to P_{i,R}$ be the composition of $\mathrm{pr}_i: \widetilde{Y} \to \widetilde{X}$ and $\mathrm{pr}: \widetilde{X} \to \mathbb{P}^n_R$. Let $P := \prod_{i=1}^m P_i$ and $P_R := \prod_{i=1}^m P_{i,R}$ (fibred products over k and R) and let $p: \widetilde{Y} \to P_R$ be $(p_1, \ldots, p_m)$. For $1 \leq i \leq m$ and $1 \leq j \leq n$ let $\partial_{i,j}$ be the derivation $x_{i,j}\partial/\partial x_{i,j}$ on P_R, where the $x_{i,j}$, $1 \leq j \leq n$, are the coordinates of the standard affine open $\mathbb{A}^n_R$ in $P_{i,R}$. Let G_i be the pullback of G to $\widetilde{X}^m$ via the projection on the ith factor. Finally, we choose a non-empty open subset $U \subset X_k$ such that G is invertible on U and the $\partial_{i,j}$ generate the tangent space at every point of U (this last condition just means that U is disjoint from the $n+1$ coordinate hyperplanes). In particular, this implies that $\mathrm{pr}: X_k \to \mathbb{P}^n_k$ is etale on U. Everything up to here will be fixed during the rest of the proof.

The following two variables σ and b will be crucial; σ is a rational number with $0 < \sigma < \varepsilon$, and b is any rational number. In the end we will first have to take σ small enough, then b large enough. For any σ and b we can find a point $x = (x_1, \ldots, x_m)$ in $X^m_k(k)$ such that:

1. $h(x_1) \geq b$ (here h denotes the height function on $X_k(\bar{k})$ associated to $\mathcal{L}$; in particular we have $h(x_1) = [k : \mathbb{Q}]^{-1} \deg(x_1^*\mathcal{L})$)
2. $h(x_2)/h(x_1) \geq 2s^2, \ldots, h(x_m)/h(x_{m-1}) \geq 2s^2$
3. $\langle x_i, x_{i+1} \rangle \geq (1 - \varepsilon/4)\|x_i\| \cdot \|x_{i+1}\|$ (here $\langle x, y \rangle$ denotes the Néron-Tate pairing in $A_k(\bar{k})$ and $\|x\|^2 = \langle x, x \rangle$)
4. $x_i \in U(k)$

(here one uses that, by the theorem of Mordell-Weil, the unit ball in $A(k) \otimes \mathbb{R}$ is compact). Let $h_i := h(x_i)$. We choose $s_i \in \mathbb{Q}$ close to $h_i^{-1/2}$: $|s_i^2 h_i - 1| < b^{-1}$ will do. Let d be a square integer which is big enough and sufficiently divisible, and put $d_i := 4ds_i^2$. Suppose that $b \geq 3$. Then $s_i/s_{i+1} \geq s$ for $1 \leq i < m$, so we can choose a section f in $\Gamma\left(Y^\circ, \mathcal{L}(\sigma - \varepsilon, s_1, \ldots, s_m)^d\right)$ as in Thm. 3.6: the index of f at x is less than σ and the norm of f at the infinite places is bounded by $\exp(c(\sigma)\sum_i d_i)$. Since the index of f at x is less than σ there exist non-negative integers $e_{i,j}$ such that $\left(\left(\prod_{i,j} \partial_{i,j}^{e_{i,j}}\right) f\right)(x) \neq 0$ and, if we write $e_i = \sum_j e_{i,j}$, $\sum_i e_i/d_i = $ (index of f at x) $< \sigma$. By Lemma 4.2 we have a non-zero integral section

$$f' = \prod_{i,j} \left(\frac{(NG_i)^{e_{i,j}}}{e_{i,j}!}\right) \left(\prod_{i,j} \left(\partial_{i,j}^{e_{i,j}}\right) f\right)(x)$$

of the metrized line bundle

$$\mathcal{M} = x^* \left(\mathcal{L}(\sigma - \varepsilon, s_1, \ldots, s_m)^d \otimes \bigotimes_{i=1}^m \mathcal{L}_i^{e_i g}\right)$$

on Spec(R). We will compute an upper bound on the degree of $\mathcal{M}$ using the properties of x and of $\mathcal{L}(\sigma - \varepsilon, s_1, \ldots, s_m)$, and a lower bound using f'. These bounds then give a contradiction.

5.2 Lemma. *There exists $c_2 \in \mathbb{R}$, not depending on b, σ and the x_i, such that*

$$[k:\mathbb{Q}]^{-1}\deg x^*\mathcal{L}(\sigma-\varepsilon, s_1,\ldots,s_m)^d \le md(\sigma-\varepsilon/2)+c_2db^{-1}$$

Proof. Let $\hat{h}$ denote the Néron-Tate height on $A_k(\bar{k})$ associated to $\mathcal{L}$. Then there exists $c_1 \in \mathbb{R}$ such that $|\hat{h}(y)-h(y)|<c_1$ and $|\langle y,z\rangle - [k:\mathbb{Q}]^{-1}\deg(y,z)^*\mathcal{P}|<c_1$ for all y, z in $A_k(k)$. Then we compute:

$$\begin{aligned}
&[k:\mathbb{Q}]^{-1}\deg x^*\mathcal{L}(\sigma-\varepsilon, s_1,\ldots,s_m)^d = \\
&= (1+\sigma-\varepsilon)ds_1^2h_1 + \sum_{i=2}^{m-1}(2+\sigma-\varepsilon)ds_i^2h_i + (1+\sigma-\varepsilon)ds_m^2h_m \\
&\quad - \sum_{i=1}^{m-1} ds_is_{i+1}[k:\mathbb{Q}]^{-1}\deg(x_i,x_{i+1})^*\mathcal{P} \\
&\le (1+\sigma-\varepsilon)d(1+b^{-1}) + \sum_{i=2}^{m-1}(2+\sigma-\varepsilon)d(1+b^{-1}) + (1+\sigma-\varepsilon)d(1+b^{-1}) \\
&\quad - \sum_{i=1}^{m-1} ds_is_{i+1}(\langle x_i,x_{i+1}\rangle - c_1) \\
&\le (m(\sigma-\varepsilon)d+2(m-1)d)\,(1+b^{-1}) - \sum_{i=1}^{m-1} ds_is_{i+1}(1-\varepsilon/4)\|x_i\|\cdot\|x_{i+1}\| + 2c_1dmb^{-1} \\
&\le (m(\sigma-\varepsilon)d+2(m-1)d)\,(1+b^{-1}) - d(1-\varepsilon/4)\sum_{i=1}^{m-1} s_is_{i+1}\sqrt{2}\hat{h}_i^{1/2}\sqrt{2}\hat{h}_{i+1}^{1/2} \\
&\quad +2c_1dmb^{-1} \\
&\le (m(\sigma-\varepsilon)d+2(m-1)d)\,(1+b^{-1}) \\
&\quad -d(1-\varepsilon/4)2\sum_{i=1}^{m-1} s_is_{i+1}(h_i-c_1)^{1/2}(h_{i+1}-c_1)^{1/2} + 2c_1dmb^{-1} \\
&\le (m(\sigma-\varepsilon)d+2(m-1)d)\,(1+b^{-1}) - 2d(1-\varepsilon/4)\sum_{i=1}^{m-1} s_ih_i^{1/2}s_{i+1}h_{i+1}^{1/2}(1-c_1b^{-1}) \\
&\quad +2c_1dmb^{-1} \\
&\le (m(\sigma-\varepsilon)d+2(m-1)d)\,(1+b^{-1}) - 2d(1-\varepsilon/4)(1-c_1b^{-1})(m-1)(1-b^{-1})^2 \\
&\quad +2c_1dmb^{-1} \\
&\le (m(\sigma-\varepsilon)d+2(m-1)d)\,(1+b^{-1}) - 2d(1-\varepsilon/4)(m-1)(1-(2+c_1)b^{-1}) + 2c_1dmb^{-1} \\
&\le md(\sigma-\varepsilon/2) + b^{-1}(2md+2dm(2+c_1)+2c_1dm) \\
&= md(\sigma-\varepsilon/2) + (6+4c_1)mdb^{-1}
\end{aligned}$$

This means that the lemma holds with $c_2 = (6+4c_1)m$. □

5.3 Lemma. *There exists $c_3 \in \mathbb{R}$, not depending on b, σ and the x_i, such that*

$$[k:\mathbb{Q}]^{-1}\deg\mathcal{M} \le md(c_3\sigma-\varepsilon/2)+c_2db^{-1}$$

Proof. We have:

$$[k:\mathbb{Q}]^{-1}\deg\mathcal{M} \le md(\sigma-\varepsilon/2)+c_2db^{-1}+\sum_{i=1}^{m}(e_i/d_i)gd_ih_i$$

$$
\begin{aligned}
&= md(\sigma - \varepsilon/2) + c_2 db^{-1} + \sum_{i=1}^{m} (e_i/d_i) g 4 d s_i^2 h_i \\
&\leq md(\sigma - \varepsilon/2) + c_2 db^{-1} + \sum_{i=1}^{m} (e_i/d_i) g 4 d(1 + b^{-1}) \\
&\leq md((1 + 8g/m)\sigma - \varepsilon/2) + c_2 db^{-1}
\end{aligned}
$$

This means that the lemma holds with $c_3 = 1 + 8g/m$. □

5.4 Lemma. *There exists $c_4(\sigma)$, not depending on b and the x_i, such that for each infinite place v of k we have $\log \|f'\|_v \leq c_4(\sigma) db^{-1}$.*

Proof. Let v be an infinite place of k. To simplify notation, we denote by $\|\cdot\|$ a norm at v. We have:

$$
\|f'\| = \left(\prod_i \|NG_i(x)\|^{e_i} \right) \cdot \left\| \left(\left(\prod_{i,j} \frac{\partial_{i,j}^{e_{i,j}}}{e_{i,j}!} \right) f \right) (x) \right\|
$$

Note that $e_i < \sigma d_i < d_i$ as $\sum_i e_i/d_i < \sigma$, and that $\sum_i d_i < 8mdb^{-1}$. The first factor of the right hand side of the formula above is easy to bound, but when it is small we will need that to bound the whole right hand side. There exists a j such that $\rho_j: \mathcal{L}(\sigma - \varepsilon, s_1, \ldots, s_m)^d \to \otimes_i \mathcal{L}_i^{d_i}$ has norm $\geq \exp(-c_5 \sum_i d_i)$ at x (this c_5 is the c_2 from Prop. 3.3). Let $h := \rho_j(f)$; then $h \in \Gamma(X_k^m, p^*\mathcal{O}(d_1, \ldots, d_m))$. The metric on $\otimes_i \mathcal{L}_i^{d_i}$ we fixed a long time ago and the pullback of the product of the standard metrics on the $\mathcal{O}(d_i)$ differ by a factor bounded by $c_6^{\sum_i d_i}$. Hence it suffices to show that there exists $c_4'(\sigma)$, such that we have:

$$
\left(\prod_i \|NG_i(x)\|^{e_i} \right) \cdot \left\| \left(\left(\prod_{i,j} \frac{\partial_{i,j}^{e_{i,j}}}{e_{i,j}!} \right) h \right) (x) \right\| \leq \exp(c_4'(\sigma) db^{-1})
$$

where the metrics are now the standard metrics. Recall that the standard metric on $\mathcal{O}(n)$ on $\mathbb{P}^m(\mathbb{C})$ is given by:

$$
\|F\|(P_0 : \cdots : P_m) = |F(P_0, \ldots, P_m)| / (\sum_i |P_i|^2)^{n/2}
$$

Let $y_i = \mathrm{pr}_i(x_i)$; then $y_i \in \mathbb{P}^n(\mathbb{C})$. Let $Z_0, \ldots, Z_n$ in $\Gamma(\mathbb{P}_k^n, \mathcal{O}(1))$ be the homogeneous coordinates of $\mathbb{P}_k^n$, and for $1 \leq j \leq n$ let $D_+(Z_j) \subset \mathbb{P}_k^n$ be the affine open subscheme given by $Z_j \neq 0$; then $D_+(Z_j) \cong \mathbb{A}_k^n$ and the Z_l/Z_j $(l \neq j)$ are coordinates on $D_+(Z_j)$. For r in $\mathbb{R}$ let $D_+(Z_j)_r$ denote the standard polydisc of radius r in $D_+(Z_j)(\mathbb{C})$ given by $|Z_l/Z_j| \leq r$ $(l \neq j)$. For each i in $\{1, \ldots, m\}$ we choose j_i in $\{0, \ldots, n\}$ such that $y_i \in D_+(Z_j)_1$, i.e., the coordinates $y_{i,j}$ of y_i in $D_+(Z_{j_i})$ satisfy $|y_{i,j}| \leq 1$. We can bound the derivative, on all standard polydiscs $D_+(Z_j)_2$ of radius 2, of G in some trivialization of $\mathcal{O}(g)$. The conclusion is that there exists $c_7 > 0$, $c_7 < 1/\|G\|_{\sup, \mathbb{P}^n(\mathbb{C})}$, such that G has no zeros in the polydiscs $\Delta_i \subset D_+(Z_{j_i})_2$ of radius $r_i := c_7 \|G\|(y_i)$ and center y_i. It follows that $\mathrm{pr}_i: X_k \to P_i$ is etale over Δ_i. Let $\Delta = \prod_{i=1}^m \Delta_i$. Then Δ is simply connected, hence $p^{-1}\Delta \subset X_k^m(\mathbb{C})$ is a finite disjoint union of copies of Δ; let Δ' denote the copy that contains x.

We trivialize $\mathcal{O}(d_i)$ on Δ_i using its section $Z_{j_i}^{d_i}$. In other words, on Δ' we write $h = h_1 \prod_i Z_{j_i}^{d_i}$. The standard norm of $Z_{j_i}^{d_i}$ on Δ_i satisfies: $(n+1)^{-d_i/2} \leq \|Z_{j_i}^{d_i}\| \leq 1$ on Δ_i. By definition we have:

$$\left\| \left(\left(\prod_{i,j} \frac{\partial_{i,j}^{e_{i,j}}}{e_{i,j}!} \right) h \right) (x) \right\| = \left| \left(\left(\prod_{i,j} \frac{\partial_{i,j}^{e_{i,j}}}{e_{i,j}!} \right) h_1 \right) (x) \right| \cdot \left(\prod_i \|Z_{j_i}^{d_i}\| \right) (x)$$

so it remains to estimate the first factor of the right hand side.

On Δ' we write $h_1 = \sum a_{(n)} z^{(n)}$, with $(n) = (n_{i,j})$ (here the $z_{i,j}$ are the standard coordinates on Δ': $z_{i,j} = \mathrm{pr}_i^*(Z_j/Z_{j_i} - y_{i,j})$). Then we have:

$$\left(\prod_{i,j} \frac{(\partial/\partial_{z_{i,j}})^{n_{i,j}}}{n_{i,j}!} \right) (h_1)(x) \;=\; a_{(n_{i,j})} \;=\; \frac{1}{(2\pi\sqrt{-1})^{mn}} \int \cdots \int_{|z_{i,j}|=r_i} h_1 \prod_{i,j} z_{i,j}^{-n_{i,j}-1} \prod_{i,j} dz_{i,j}$$

Hence for (n) with $\sum_j n_{i,j} = e_i$ for all i we have $|a_{(n_{i,j})}| \leq \|h_1\|_{\sup,\Delta'} \prod_i r_i^{-e_i}$. Now note that $\|h_1\|_{\sup,\Delta'}$ is bounded above by some $\exp(c_4''(\sigma)db^{-1})$ by the arguments above plus the fact that the norm of f is bounded by $\exp(c(\sigma)\sum_i d_i)$, hence we have:

$$\left| \left(\prod_{i,j} \frac{(\partial/\partial_{z_{i,j}})^{n_{i,j}}}{n_{i,j}!} \right) (h_1)(x) \right| \leq \exp(c_4''(\sigma)db^{-1}) \prod_i r_i^{-e_i}$$

A standard computation shows that on Δ_i one can write $\partial_{i,j} = \sum_l b_{i,j,l}(\partial/\partial_{z_{i,l}})$ with $|b_{i,j,l}(y_i)| \leq 1$ (in fact, $b_{i,j,l}(y_i) \in \{0, \pm y_{i,l}\}$). Using this, we can write:

$$\begin{aligned} \left| \left(\left(\prod_{i,j} \frac{\partial_{i,j}^{e_{i,j}}}{e_{i,j}!} \right) h_1 \right) (x) \right| &= \left| \sum_{(n_{i,j})} C_{(n_{i,j})}(x) \left(\prod_{i,j} \frac{(\partial/\partial_{z_{i,j}})^{n_{i,j}}}{n_{i,j}!} \right) (h_1)(x) \right| \\ &= \left| \sum_{(n_{i,j})} C_{(n_{i,j})}(x) \cdot a_{(n_{i,j})} \right| \end{aligned}$$

where the $C_{(n_{i,j})}$ are certain polynomials in the $b_{i,j,l}$. Note that $C_{(n_{i,j})}(x) = 0$ unless for all i one has $\sum_j n_{i,j} = e_i$. From $|b_{i,j,l}(x)| \leq 1$, and some combinatorics, it follows that:

$$\left| C_{(n_{i,j})}(x) \right| \leq \prod_i \frac{e_i!}{e_{i,1}! \cdots e_{i,n}!}$$

Note that the right hand side does not depend on $(n_{i,j})$. Using this one gets:

$$\sum_{(n_{i,j})} |C_{(n_{i,j})}(x)| \leq \prod_i \frac{e_i!}{e_{i,1}! \cdots e_{i,n}!} \frac{(e_i + n - 1)!}{(n-1)! e_i!} \leq \prod_i (2^{e_i+n-1} n^{e_i}) \leq 2^{m(n-1)} (2n)^{\sum_i \sigma d_i}$$

This means that finally we have:

$$\left| \left(\left(\prod_{i,j} \frac{\partial_{i,j}^{e_{i,j}}}{e_{i,j}!} \right) h_1 \right) (x) \right| \leq \exp(c_4''(\sigma)db^{-1}) \cdot \left(\prod_i r_i^{-e_i} \right) \cdot 2^{m(n-1)} (2n)^{4md/b}$$

which finishes the proof. $\square$

5.5 Lemma. *There exists $c_8(\sigma)$, not depending on b and the x_i, such that*

$$[k:\mathbb{Q}]^{-1}\deg\mathcal{M} \;\geq\; -c_8(\sigma)db^{-1}$$

Proof. By definition we have:

$$\deg\mathcal{M} \;=\; \log\#\left(\frac{\mathcal{M}}{R\cdot f'}\right) - \sum_{v|\infty}\epsilon(v)\log\|f'\|_v \;\geq\; -[k:\mathbb{Q}]c_4(\sigma)db^{-1}$$

□

We can now finish the proof of Thm. 5.1. Take $\sigma < \varepsilon/(2c_3)$ (c_3 as in Lemma 5.3). Then for b big enough, Lemma 5.3 and Lemma 5.5 give contradicting estimates for $[k:\mathbb{Q}]^{-1}\deg\mathcal{M}$. □

Chapter XII

Points of Degree d on Curves over Number Fields

by Gerard van der Geer

Arithmetic questions on the number of points of degree d on a smooth (irreducible) algebraic curve over a number field lead to geometric questions about the curve by using Faltings's big theorem. We discuss here some questions and conjectures initiated by Abramovich and Harris in [4]. We refer to the end of this paper for a discussion of the recent literature.

Faltings's theorem (previously Mordell's conjecture) says that an irreducible curve C of geometric genus $g \geq 2$ defined over a number field K possesses only finitely many K-rational points. But since over an algebraic closure $\overline{K}$ of K the number of points has the same cardinality as $\overline{K}$ itself one is led to the question how the number of points grows with the degree of the extension L/K over which we consider our curve. More precisely, what can we say about

$$\Gamma_d(C, K) = \{p \in C(\overline{K}) : [K(p) : K] \leq d\},$$

the set of points of degree $\leq d$?

A first question is: *is it finite ?* The answer is: no, not always. For if we take a curve C which admits a non-constant morphism $C \to \mathbb{P}^1$ defined over K of degree $\leq d$ then we have $\#\Gamma_d(C, K) = \infty$ in a trivial way: over every rational point on $\mathbb{P}^1$ there lies a point of degree $\leq d$. Similarly, if C admits a non-constant K-morphism $C \to E$, where E is an elliptic curve with $\#E(K) = \infty$ (or equivalently, with K-rank ≥ 1) we find infinitely many points of degree $\leq d$.

This raises a multitude of questions. To begin with, the question arises whether $\#\Gamma_d(C, K)$ is finite when C does not admit a dominant K-morphism $C \to \mathbb{P}^1$ or $C \to E$ of degree $\leq d$; or phrased differently, if $\#\Gamma_d(C, L) = \infty$ for some finite extension L/K, does this imply that C admits a dominant morphism of degree $\leq d$ to $\mathbb{P}^1$ or to an elliptic curve ?

As we shall see, the answer is : no! There are curves with infinitely many "interesting" points of degree d (i.e., not obtained from a morphism of degree d to $\mathbb{P}^1$ or to an elliptic curve).

To put the questions in the framework of this volume we introduce the algebraic variety (for $d \in \mathbb{Z}_{\geq 1}$)

$$C^{(d)} = \mathrm{Sym}^{(d)}(C) = C^d/S_d,$$

the d-fold symmetric product of our curve. We have morphisms

$$\phi_d : C^{(d)} \to \mathrm{Pic}^{(d)}(C) \quad \text{given on geometric points by } \{p_1, \ldots, p_d\} \mapsto \sum_{i=1}^{d} p_i.$$

Here $\mathrm{Pic}^{(d)}(C)$ is the variety of divisor classes of degree d. The (geometric) fibres of this morphism are the linear systems:

$$|D| = \left\{ (p_1, \ldots, p_d) : \sum p_i \sim D \right\},$$

where $\sim$ denotes linear equivalence.

Some remarks are in order here.

(i) By Riemann-Roch we have $h^0(D) \geq 1$ for $d \geq g$, hence ϕ_d is surjective.

(ii) If C does not admit a dominant morphism $C \to \mathbb{P}^1$ of degree $\leq d$ then ϕ_d is injective.

(iii) Any curve of genus g admits a morphism of degree $\leq [(g+3)/2]$ onto $\mathbb{P}^1$.

We need more notation. Define

$$W_d^r = \{D \in \mathrm{Pic}^{(d)}(C) : h^0(D) \geq r+1\}$$

and set

$$W_d = W_d^0 = \phi_d(C^{(d)}).$$

These (functors) are (represented by) algebraic varieties defined over K. Examples of these are:

$$W_{g-1}^0 = \Theta \subset \mathrm{Pic}^{(g-1)}(C),$$

the famous theta divisor of effective divisor classes of degree $g-1$ on C and

$$W_{g-1}^1 = \mathrm{Sing}(\Theta),$$

the singular locus of the theta divisor. We know that $\dim \mathrm{Sing}(\Theta) \geq g-4$.

Let us now assume that C does not admit a dominant morphism to $\mathbb{P}^1$ of degree $\leq d$. Then ϕ_d is injective. A point of C of degree d (over K) determines a K-rational point of $C^{(d)}$ (note that the natural local coordinates near $\{p_1, \ldots, p_d\}$ are the elementary symmetric functions in local parameters t_i near p_i) and in turn this determines a K-rational point of W_d. So if we know that W_d contains only finitely many K-rational points then necessarily $\Gamma_d(C, K)$ is finite. But here Faltings's Theorem on rational points on subvarieties of abelian varieties comes in and weaves the present theme into the texture of this volume:

> *if W_d does not contain the translate of a positive dimensional abelian variety then $W_d(K)$ is finite, and then $\Gamma_d(C, K)$ is finite too.*

All this leads us to the question: when does a jacobian variety contain an abelian subvariety (always assumed to be of positive dimension) in its W_d ? One way for this to happen is when C is a covering of a (complete irreducible smooth) curve D. If $\pi: C \to D$ is a morphism of degree n with $g(D) = h$ we have an induced morphism

$$\pi^*: \mathrm{Pic}^{(h)}(D) \to \mathrm{Pic}^{(nh)}(C)$$

whose image lands in W_{nh} (since $\phi_d(D^{(h)}) = \mathrm{Pic}^{(h)}(D)$) and thus we see that for $nh \leq d$ we find a translate of an abelian variety in $W_d(C)$. This could also happen if our C is the image of a curve C' which is a d-fold covering of a curve D. Again we are led to speculate about the converse:

1 Question. If C is a curve of genus g and $W_d(C)$ for some $d < g$ contains a *maximal* abelian subvariety of dimension h then does it follow that C is the image of a curve C' which admits a dominant morphism $C' \to C''$ of degree $\leq d/h$ with $g(C'') = h$? □

Let us call this question $A(d, h; g)$, i.e., does it hold for all C when d, h, g are fixed. A related question is:

2 Question. If C admits a dominant morphism $C' \to C$ and C' admits a dominant morphism $C' \to C''$ of degree $\leq d/h$ and $g(C'') = h$, then does C itself admit a dominant morphism $C \to C'''$ of degree $\leq d/h$ with $g(C''') \leq h$? □

Call this question $S(d, h; g)$, i.e., does it hold for all C of genus g and the given values of d and h ?

For points of degree d we have the relevant question:

3 Question. Is it true for all irreducible smooth curves of genus g over a number field K that $\#\Gamma_d(C, L) = \infty$ for some finite extension L/K if and only if C admits a map of degree $\leq d$ to $\mathbb{P}^1$ or to an elliptic curve ? □

Call this question $F(d, g)$. Note that positive answers to $A(d, h; g)$ and $S(d, h; g)$ for all h with $1 \leq h \leq d$ imply $F(d, g)$ using Faltings's theorem. Indeed, we may deduce from it that C admits a morphism of degree $\leq d/h$ to a curve C' of genus $\leq h$. Either C' is of genus ≤ 1 or ≥ 2. In the latter case the curve C' admits a morphism of degree $\leq h$ to $\mathbb{P}^1$ since $h \geq [(h+3)/2]$. In any case we find a morphism of degree $\leq d$ to $\mathbb{P}^1$ or to an elliptic curve. We do not need consider all h in the range $1, \ldots, g$ in view of the following remark.

4 Remark. An affirmative answer to $A(d, h; g)$ implies the following statement:

> *for $d < g$ the variety W_d cannot contain an abelian variety of dimension $> d/2$.*

This last statement was proved to be true by Debarre and Fahlaoui [16]. □

There are some scattered results concerning question $F(d, g)$. Abramovich and Harris proved it in [4] for $d = 2$ and 3 (all g) and for $d = 4$ provided that in the latter case $g \neq 7$. Using the same arguments one can also show it for other d provided g does not lie in a certain interval, e.g. $d = 6$ and $g \leq 10$ or $g \geq 17$. One could also consider a sharpened version of the question : suppose that $\#\Gamma_d(C, L) = \infty$ for some finite extension L, but $\#\Gamma_e(C, M) < \infty$ for all $e < d$ and all finite extensions M/K, then does it follow that C admits a non-constant morphism of degree d to $\mathbb{P}^1$ or to an elliptic curve ? It seems that this sharpened question admits a positive answer for $d = p$, a prime and $g \leq 2p - 2$ or $g \geq \binom{p}{2} + 2$. None of the questions $A(d, h; g)$ and $S(d, h; g)$ has a positive answer for all triples for which the question makes sense. Abramovich and Harris conjectured in [4] that $F(d, g)$ is true for all tuples (d, g) with $d \geq 1$ and $g \geq 0$. However, this was disproved shortly afterwards by Debarre and Fahlaoui [16].

The questions $A(d, h; g)$ and $S(d, h; g)$ do not admit a positive answer for many triples (d, h, g). To point out some positive statements, one can see that $A(d, h; g)$ has an affirmative answer for $h = 1$. Similarly, the question $S(d, h; g)$ has an affirmative answer for $g = 0$. Coppens [15] proved that $S(d, h, g)$ holds for d a prime and g large using Castelnuovo theory. He also shows that counterexamples to $S(d, h, g)$ produce counterexamples to $S(md, h, g)$ for large enough g.

In the following we give a few principles from [16] which give rise to some affirmative answers to the questions A and S and we give an easy counterexample to $A(2h, h, 2h + 1)$. For more results we refer to the literature, although the terrain is largely unexplored and the interested reader might find rewarding challenges there.

5 Lemma. *Assume that $\Theta \subset \mathrm{Pic}^{(g-1)}(C)$ contains a subvariety Z stable under translation by an abelian subvariety $A \subsetneq \mathrm{Jac}(C)$. Then*

$$\dim(Z) + \dim(A) \leq g - 1.$$

Proof. We may assume that Z is irreducible and meets Θ_{reg} (otherwise, replace Z by $Z + W_d(C) - W_d(C)$, where $d + 1 =$ multiplicity of Θ in the generic point of Z). Consider the Gauss map

$$g\colon \Theta_{\mathrm{reg}} \to \mathbb{P}(T_0(\mathrm{Jac}(C))^*), \quad x \mapsto T_{\Theta,x}.$$

If $x \in Z \cap \Theta_{\mathrm{reg}}$, then $T_{\Theta,x}$ contains the fixed space $x + T_{A,0}$, hence we find a map

$$g\colon Z \cap \Theta_{\mathrm{reg}} \to \mathbb{P}((T_{\mathrm{Jac},0}/T_{A,0})^*).$$

But for a Jacobian the Gauss map has finite fibres (see e.g. [5]), hence

$$\dim(Z) \leq g - \dim(A) - 1.$$

□

Similarly, if W_d^r contains Z which is stable by A and if $d \leq g - 1 + r$ then

$$\dim(Z) + \dim(A) \leq d - 2r.$$

(Apply the Lemma to $Z + W_{g-1-d+r} - W_r \subseteq W_{g-1} = \Theta$.)
When do we have equality ?

6 Lemma. *Suppose that C is such that $W_d(C)$ contains Z stable under translation by an abelian variety $A \neq (0)$ in $\mathrm{Jac}(C)$. Assume that $\dim(Z) + \dim(A) = d$ and $\dim(Z) + d \leq g - 1$. Then there exists a curve B of genus $h = \dim(A)$ and a morphism $\pi\colon C \to B$ of degree 2 such that $A \cong \mathrm{Jac}(B)$ and $Z = \pi^*(\mathrm{Pic}^{(h)}(B)) + W_{d-2h}(C)$.*

There is a variation of this Lemma in which one reads W_d^r instead of W_d and assumes $\dim(Z) + \dim(A) = d - 2r$ and in which one then finds that $Z = \pi^*\mathrm{Pic}^{(h+r)}(B) + W_{d-2r-2h}(C)$. Taking $Z = A$ one obtains the corollary:

7 Corollary. *Let C be such that $W_d^r(C)$ contains a translate of an abelian variety A of dimension $h > 0$. Assume that $d \leq g - 1 + r$. Then $\dim(A) \leq d/2 - r$. If $d \leq \frac{2}{3}(g-1+r)$ we have equality if and only if d is even and if there exists a curve B of genus $d/2 - r$ and a morphism $\pi: C \to B$ of degree 2 such that $A = \pi^*(\mathrm{Pic}^{d/2}(B))$.*

Note that this implies the validity of $A(2h, h; g)$ for $h < g/3$.

It is not difficult to construct counterexamples to $A(2h, h, 2h+1)$ for $h \geq 4$. One uses Prym varieties of double covers as follows.

Let $C \to D$ be a double etale cover of a smooth irreducible curve of genus $g(D) = h+1$. Then the genus $g(C)$ equals $2h+1$ and we can consider the Prym variety

$$P = \mathrm{Prym}(C \to D) = \ker\{\mathrm{Nm} : \mathrm{Jac}(C) \to \mathrm{Jac}(D)\}^0.$$

Here Nm is the norm map. This connected component of this kernel has dimension h and can (up to translation) be identified with

$$\{d \in W_{g-1=2h}(C) : \pi_*(d) = K_D, h^0(d) \equiv 0(\mathrm{mod} 2)\},$$

where K_D is the canonical divisor class of D. The claim is then that for general D and $h \geq 4$ there does not exist a double covering $p: C' \to B$ with $g(B) = h$ and a non-constant morphism $q: C' \to C$. Indeed, suppose that it does exist. Then we find $q_*(p^*(\mathrm{Jac}(B)))$ inside $\mathrm{Jac}(C)$. For D general it is known that the jacobian $\mathrm{Jac}(D)$ is simple and that the Prym P is not isogenous to a jacobian for $h \geq 4$. We see that $\mathrm{Jac}(C)$ is isogenous to a product $P \times \mathrm{Jac}(D)$, both factors of which are simple and not isogenous to each other. It follows that $q_*(p^*(\mathrm{Jac}(B)))$ is a point. But this is impossible because then for all $D \in \mathrm{Jac}(B)$ we have $\deg(q)p^*(D) = q^*q_*p^*(D) = 0$ which contradicts the fact that the kernel of p^* is finite.

8 Additional Remarks

If we change the point of view somewhat (from the jacobian to the abelian variety in its $W_d(C)$) we might pose the question differently: which curves do lie on a given abelian variety ? Answers to this type of questions (very interesting in their own right) can also be helpful in that they put restrictions on the possible abelian varieties.

Instead of curves over a number field one might consider curves over function fields and ask these questions there. Questions A and S make sense for curves over any field.

9 Discussion of the Literature

Many of the questions treated here were first stated explicitly in a paper of Abramovich and Harris [4] (the title of which is not what you would think). The merits of this interesting paper lie more in the questions it poses than in the answers it provides. The questions "A", "S" and "F" were raised there (though in a slightly different form and question "F" was stated as a conjecture there). The thesis of Abramovich [1] contains additional material. The first author wrote a corrigendum [3] to [4] (the title of which is not what he thinks) pointing out a number of gaps and misprints in [4], e.g., the proof of Lemma 6. In Lemma 7 one should read : $r_{k+1} - r_k \geq r_k - r_{k-1}$. Besides that there are more lapses. In particular, Lemma 8 (p. 223–224) of [4] is false. Abramovich tells me that he manages to salvage Theorem 2 with a lot of effort, but so far he did not publish how he did that.

The main conjecture of [4] was disproved by Debarre and Fahlaoui [16]. They give partial answers to the questions $A(d,h;g)$ and $S(d,h;g)$.

Debarre and Klassen [17] classify all points of degree d on a smooth plane curve of degree d.

In the paper [6] of Alzati and Pirola the reader will find some related results. Finally, in the paper by Vojta [84] the reader will find a different approach to points of given degree over a number field using arithmetic surfaces.

Chapter XIII

"The" General Case of S. Lang's Conjecture (after Faltings)

by Frans Oort

In this talk we discuss (see Thm. 3.2 below) the main result of [23]. It proves the conjecture stated in [37], page 321, lines 12–15, and it generalizes the main result of [22], which is the theme of this conference. Also see [81], Theorem 0.3, and §10.

We assume (for simplicity) that $\mathbb{Q} \subset K \subset k = \overline{k}$, i.e., K is a field of characteristic zero contained in an algebraically closed field k.

1 The Special Subset of a Variety

1.1 Construction. Let X be a variety over an algebraically closed field k. Consider all abelian varieties C, and all non-constant rational maps

$$f : C \cdots \rightarrow X.$$

Let $\mathrm{Sp}(X) \subset X$ be the Zariski closure of all the images of these maps. This is called the **special subset** of X. (Note that a rational map does not have an "image" in general, but f is defined on a non-empty open set, and the closure of that image is well-defined, etc.) If Y is a (reducible) algebraic set, $Y = \bigcup_i X_i$, then we write $\mathrm{Sp}(Y) := \bigcup_i \mathrm{Sp}(X_i)$. □

1.2 Question. Do we really need to take the Zariski closure, or is just the union of all images already closed? □

This seems to be unknown in general, but in the case considered in the next section this is true.

1.3 Example. In the situation of [22] we have $\mathrm{Sp}(X) = \emptyset$, and Thm. 3.1 and Thm. 3.2 below generalize the main result of [22]. □

Note that an elliptic curve can be mapped onto $\mathbb{P}^1$, hence we see that $\mathrm{Sp}(X)$ contains all rational curves contained in X. In fact, for every group variety G and for every non-constant rational map f from G into X the image of f is contained in $\mathrm{Sp}(X)$, and $\mathrm{Sp}(X)$ could be defined using all such maps.

1.4 The Kodaira Dimension

For a variety X over k one can define the Kodaira dimension, denoted by $\kappa(X)$ (see [79], §6). We briefly indicate the idea: take a complete normal model of X, let K_X be its canonical divisor, let n be a rational integer. If every section in the sheaf $\mathcal{L}(nK_X)$ is zero (for every positive rational integer n) we write $\kappa(X) = -\infty$. In all other cases we write $\kappa(X)$ for the maximum (for all n) of the dimensions of the image of the (multicanonical) rational map defined by nK_X. One can show that this is a birational invariant.

We say that X is **of general type** (or: of hyperbolic type) if

$$\dim(X) = \kappa(X).$$

Some examples:
(d=1) For complete algebraic curves we have:

$$\begin{aligned} g(C) = 0 &\iff \kappa(C) = -\infty \\ g(E) = 1 &\iff \kappa(E) = 0 \\ g(C) \geq 2 &\iff \kappa(C) = 1 \iff C \text{ is of general type} \end{aligned}$$

We have: $\kappa(\mathbb{P}^d) = -\infty$ for $d > 0$, $\kappa(\mathbb{P}^0) = 0$ and $\kappa(\text{abelian variety}) = 0$.

1.5 Conjecture. *(S. Lang, cf. [38], page 17, 3.5)* **(GT)**:

$$\mathrm{Sp}(X) \neq X \iff X \textit{ is of general type.}$$

1.6 Remarks. If X is defined over K, then $\mathrm{Sp}(X_k)$ is also defined over K, i.e., there exists a K-closed subset $\mathrm{Sp}(X) \subset X$ so that $\mathrm{Sp}(X)_k = \mathrm{Sp}(X_k)$ (but note that the abelian varieties C and rational maps $f : C \cdots \to X_k$ need not be defined over K).

For algebraic surfaces not of general type one sees that indeed $\mathrm{Sp}(X) = X$; for K3-surfaces, see [51], page 351. It seems that Conjecture 1.5 for surfaces of general type is still open. □

1.7 Conjecture (Lang's conjecture). (LC): *Let X be defined over K, suppose that K is of finite type over $\mathbb{Q}$, then*

$$\#\{x \mid x \in X(K) \text{ and } x \notin \mathrm{Sp}(X_k)\} < \infty$$

(cf. [38], page 17, 3.6). For **(GT)** and **(LC)** also see [39], page 191.

Note that for a curve X the special subset is empty iff the genus $g(X) \geq 2$. Hence **(GT)** is easily proved for curves, and **(LC)** is "Mordell's conjecture", proved by Faltings (first for number fields, later for fields of finite type over $\mathbb{Q}$, cf. [24], page 205, Theorem 3).

If the conjectures **(GT)** and **(LC)** are true we could conclude that the following conjecture holds:

1.8 Conjecture. (cf. [80], page 46) *If X is a variety defined over K, a field of finite type over $\mathbb{Q}$, and X is of general type, then the Zariski closure in X of $X(K)$ is a proper subset of X.*

(This conjecture was mentioned by Bombieri in 1980, cf. [55], page 208).

1.9 Remark. Note that the definition of $\mathrm{Sp}(X)$ is purely "geometric", but we shall see the importance of this notion for arithmetic questions. □

2 The Special Subset of a Subvariety of an Abelian Variety

We denote by A an abelian variety over k, and by $X \subset A$ a closed subvariety. Notation:

$$Z(X) = \{x \in X \mid \exists B \subset A, \text{ an abelian variety}, \dim B > 0, \text{ and } x + B \subset X\}.$$

Note that $Z(X) \subset \mathrm{Sp}(X)$. From the fact that any rational map between abelian varieties extends to a morphism of varieties (see [14], Chapter V, Thm. 3.1), and moreover that any morphism of varieties between abelian varieties is, up to translation, a homomorphism of abelian varieties (see [14], Chapter V, Cor. 3.6), it follows that $\mathrm{Sp}(X)$ is the Zariski closure of $Z(X)$.

2.1 Theorem. (Ueno; Kawamata [32]; Abramovich [1], Thms. 1,2) $Z(X) = \mathrm{Sp}(X)$, *in particular* $Z(X)$ *is closed in* X. *For every component* Z_i *of* $Z(X)$ *we write* $B_i := \mathrm{Stab}(Z_i)^0$; *then we have* $\dim(B_i) > 0$.

2.2 Lemma. (cf. [54], Lemma 1.5; [22], proof of Lemma 4.1; [1], 1.2.2, Lemma 3) *Let* A *be an abelian variety over* k, *let* $W \subset A$ *be a closed subvariety (in particular* W *is irreducible and reduced), and let* $t \in \mathbb{Z}_{>1}$. *Suppose that* $tW = W$. *Then* W *is a translate of an abelian subvariety of* A.

(Comment: by $tW = W$ we mean that the map $[t]: A \to A$ maps the subvariety W onto itself; note that this map is finite, and hence it suffices to require that for all w in W we have $t \cdot w \in W$; in that case $tW \subset W$, and we conclude $tW = W$ by remarking $\dim(tW) = \dim W$.)

2.3 Sketch of a Proof of Theorem 2.1

Using Lemma 2.2 we sketch a proof of that theorem. We choose an integer $q \in \mathbb{Z}_{>1}$. Following Faltings we define for every $m > 1$ the map

$$F_m: X^m \longrightarrow A^{m-1}$$

by:

$$F_m(a_1, \ldots, a_m) := (qa_1 - a_2, qa_2 - a_3, \ldots, qa_{m-1} - a_m).$$

Define Y'_m by the cartesian square (i.e., pull-back diagram):

$$\begin{array}{ccc} X^m & \longrightarrow & A^{m-1} \\ \uparrow & & \uparrow {\scriptstyle g=(q-1)\cdot\Delta} \\ Y'_m & \xrightarrow{h} & X \end{array}$$

where $g(x) := ((q-1)x, \ldots, (q-1)x)$. Let $Y_m \subset A \times X$ be the image of

$$(\mathrm{proj}_1, h): Y'_m \to A \times X.$$

Note:

$$(a, x) \in Y'_m \text{ and } a_1 = x + b' \iff a_i = x + q^{i-1}b' \text{ for } 1 \leq i \leq m.$$

Note that

$$(x+b',x)\in Y_m \iff (x+b',x+qb',\ldots,x+q^{m-1}b')\in Y'_m,$$
$$\iff x+b'\in X,\ldots,x+q^{m-1}b'\in X.$$

Suppose $B'\subset X$ is closed, $x\in X$, and $b'\in B'$ with $x+b'\in X$; then

$$qB'=B'\Rightarrow (x+b',x)\in Y_m;$$

we are going to try to prove the "opposite implication". We observe:
a) the maps $(\mathrm{proj}_1,h)\colon Y'_m\to Y_m\subset A\times X$ make $Y_{m+1}\subset Y_m\subset A\times X$ into a descending chain $(2\le m)$ of closed subschemes; clearly this is stationary, let n be chosen such that $Y_m=Y_n=:Y$ for all $m\ge n$;
b) for $x\in X$ consider the fiber

$$Y(x):=h^{-1}(x)\subset A\times\{x\}\cong A$$

(as a closed set of A); note that $((x,\ldots,x),x)\in Y'_m$, hence $x\in Y(x)$; note that $Y=Y_m=Y_{m+1}$ (for $m\ge n$) implies

$$(x+b,x)\in Y_m\iff (x+b,x)\in Y_{m+1}\Longrightarrow (x+qb,x)\in Y_m;$$

we write $B'':=-x+Y(x)$, and we see that multiplication by q maps B'' into itself, hence a power $t=q^{?}>1$ of it maps some highest dimensional irreducible component B' of B'' onto itself; now we apply Lemma 2.2 with $W=B'$; we see that B' is a translate of an abelian subvariety B of A;
c) hence $Z(X)=\{x\in X\mid \dim(Y(x))>0\}$; the projection map $Y\to X$ is proper, so by general theory we conclude that $Z(X)$ is a closed subset of X; the equality $Z(X)=\mathrm{Sp}(X)$ has now been proved;
d) let Z_i be an irreducible component of $Z(X)$ and let x_i be its generic point; the method of step (b) (with some minor changes due to the fact that $k_i:=\overline{k(x_i)}\neq k$) applied to x_i gives a non-zero abelian subvariety B'_i of $A\otimes_k k_i$; by rigidity of abelian subvarieties (cf. [40], page 26, Thm. 5) B'_i is defined over k, i.e., we have $B_i\subset A$ with $B'_i=B_i\otimes_k k_i$; by construction, B_i stabilizes Z_i.

2.4 The Ueno Fibration

Let $X\subset A$ over an algebraically closed field k, and let B be the connected component of the stabilizer of X:

$$B:=\{a\in A\mid a+X\subset X\}^0_{\mathrm{red}}.$$

This is an abelian subvariety of A. One can consider the quotient X/B, i.e., the image of X under the mapping $A\to A/B$. The mapping $X\to X/B$ is called the Ueno-fibration of X (e.g. cf. [79], pp. 120/121). The Kodaira dimension of X equals $\dim(X/B)$ (cf. [79], Thm. 10.9), hence conjecture **(GT)** (see §1) holds for subvarieties of abelian varieties, also cf. [32], Thm 4. For connections with a conjecture made by A. Bloch in 1926, cf. [57] and [32]. For $Z(X)$, when X is contained in a semi-abelian variety, cf. [56], Lemma 4.1. Also see [2], Thm. 1.

3 The Arithmetic Case

In this section we suppose that k is an algebraic closure of K. The crucial result is:

3.1 Theorem. (cf. [23], Thm. 4.1, and §5) *Suppose K is of finite type over $\mathbb{Q}$, and A is an abelian variety over K, and $X \subset A$ is a K-closed subset; then*

$$\#\{x \mid x \in X(K) \text{ and } x \notin \mathrm{Sp}(X_k)\} < \infty.$$

(If $x \in X$ then $x \otimes k \in X_k$, but this point we still denote by x. This theorem and its proof are generalizations of results and methods from [22].)

3.2 Theorem. (cf. [23], Thm. 4.2, notations and assumptions as in the previous theorem) *(Either $X(K)$ is empty, or) there exist $c_i \in X(K)$ and abelian subvarieties $C_i \subset A$ (defined over K) such that*

$$X(K) = \bigcup_{j=1}^{m} (c_j + C_j(K)).$$

In other words, the irreducible components of the Zariski closure $\overline{X(K)}$ of $X(K)$ in X are translates of abelian subvarieties of A.

Proof. Induction on $\dim(X)$. For $\dim(X) = 0$ it is trivial. If $X(K)$ is finite we are done. Let T be a positive dimensional irreducible component of $\overline{X(K)}$. Note that T is geometrically irreducible since $T(K)$ is Zariski dense in T. From Thm. 3.1 it follows that $T_k = \mathrm{Sp}(T_k)$. Let $B := \mathrm{Stab}(T)^0_{\mathrm{red}}$; note that $(\mathrm{Stab}(T_k))^0 = ((\mathrm{Stab}(T)^0)_k$; hence B_k is a non-zero abelian subvariety of A_k by Thm. 2.1, and we see that B is an abelian subvariety of A. By induction, T/B is a translate of an abelian subvariety C/B of A/B, where C is an abelian subvariety of A containing B. It follows that T is a translate of C. □

3.3 Remarks. This theorem proves the conjecture by Lang from 1960, cf. [37], page 29, which is a special case of **(LC)**.

It does happen that the dimension of the Zariski closure of $X(K)$ depends on the choice of K (and hence I do not agree with "...that the higher dimensional part of the closure of X(number field) should be geometric; i.e., independent of that number field", cf. [81], page 22, lines 11–12). As an easy example one can take $X = A = E$, an elliptic curve which has only finitely many points rational over K. One can remark that the top dimensional part is not geometric, but its dimension eventually is.

Note that Thm. 3.2 does not generalize directly to semi-abelian varieties (e.g., take an elliptic curve $E \subset \mathbb{P}^2$ over a number field K such that $E(K)$ is not finite, and remove the 3 coordinate axes, obtaining $U \subset \mathbb{G}_m \times \mathbb{G}_m$; one can also take a singular rational curve or a conic instead of E). □

4 Related Conjectures and Results

4.1 Integral Points

Suppose S is a finite set of discrete valuations on a number field K, and let $R := \mathcal{O}_S$ be the integers outside S. One can consider points "over R". For curves it turns out to be essential that the curve has only finitely many automorphisms (this condition became natural in the Kodaira-Parshin construction), and now we have a better understanding of theorems of Siegel and Mahler: if C is either $\mathbb{P}^1_R - \{0, 1, \infty\}$, or $E - \{0\}$ (where E is an elliptic curve over R), or C is a curve over K of genus at least two, then $C(R)$ is finite (cf. Siegel [74], Mahler [44], Faltings [22]; see the discussion on page 319 of [37]; for the first case see [12]). For a generalization, see [19].

In [23] Faltings gives an example that one cannot hope for finiteness of integral points on an open set in an abelian varieties (in case the open set is not affine).

See [36], page 219 for a conjecture concerning finiteness of number of integral points on an affine open set in an abelian variety.

See [83], Thm. 0.2 for the statement of a result which generalizes this conjecture: take R as above, take a closed subscheme $\mathcal{X}$ of a model $\mathcal{A}$ over R of a semi-abelian variety A over K; the set of integral points on $\mathcal{X}$ is contained in a finite number of translates of semi-abelian subvarieties (Definition: A group variety A is called a **semi-abelian variety** if it contains a linear group $L \subset A$ such that L is an algebraic torus (i.e., product of copies of $\mathbb{G}_m$ over the algebraic closure of the field of definition), such that A/L is an abelian variety, i.e., if A is the extension over an abelian variety by an algebraic torus). We see, cf. [83], Coroll. 0.5, that on a semi-abelian variety, in the complement of an ample divisor we have only finitely many integral points.

4.2 Torsion Points on Subvarieties

Manin and Mumford studied a question which was settled by Raynaud:

4.3 Theorem. (Manin-Mumford conjecture, proved by Raynaud, cf. [60], Thm. 1) *Let A be a complex torus (i.e., $\mathbb{C}^g$ modulo a lattice), and let $S \subset A$ be the image of a Riemann surface into A. Suppose that S is not the Riemann surface of an elliptic curve (i.e., not the translate of a subtorus). Then the set of torsion points of A contained in S is finite.*

More generally one can consider a subgroup of finite type (i.e., finitely generated as $\mathbb{Z}$-module, or take it of "finite rank", see below) inside an abelian variety and intersect with a subvariety (cf. [45], translation page 189, see [36], page 221).

We see that [38], page 37, Conjecture 6.3 can now be derived from the existing literature (Liardet, Laurent [41], Hindry [29]: see the discussion on pp. 37–39 of [38], use Faltings [23]):

> *Consider a semi-abelian variety A over $\mathbb{C}$, a finitely generated subgroup $\Gamma_0 \subset A(\mathbb{C})$, and a subvariety $X \subset A$. Let*
>
> $$\Gamma := \{a \in A(\mathbb{C}) \mid \exists n \in \mathbb{Z}_{>0} \text{ with } n \cdot a \in \Gamma_0\}.$$
>
> *Then X contains a finite number $b_i + B_i$ of translates of semi-abelian subvarieties of A such that*
>
> $$X(\mathbb{C}) \cap \Gamma \subset \bigcup (b_i + B_i(\mathbb{C})) \subset X.$$

Bibliography

[1] D. Abramovich, *Subvarieties of abelian varieties and Jacobians of curves*, PhD-Thesis, Harvard University, April 1991.

[2] D. Abramovich, *Subvarieties of semiabelian varieties*, Preprint, March 1992.

[3] D. Abramovich, *Addendum to "Curves and abelian varieties on $W_d(C)$"*, (sic) December 1991.

[4] D. Abramovich, J. Harris, *Abelian varieties and curves in $W_d(C)$*, Compositio Math. 78, 227–238 (1991).

[5] E. Arbarello, M. Cornalba, P.A. Griffith and J. Harris, *Geometry of Algebraic Curves, Vol. I*, Grundlehren der Mathematische Wissenschaften 267, Springer.

[6] A. Alzati and G. Pirola, *On curves in $C^{(2)}$ generating proper abelian subvarieties of $J(C)$*, Preprint 1992.

[7] A. Baker, *The theory of linear forms in logarithms*, In: Transcendence Theory: Advances and Applications, Academic Press, 1977, pp. 1–27.

[8] A. Baker, *Transcendental Number Theory*, Cambridge University Press, 1975.

[9] E. Bombieri, *The Mordell Conjecture revisited*, Ann. Scu. Sup. Pisa (1991), 615–640.

[10] E. Bombieri and W.M. Schmidt, *On Thue's equation*, Inv. Math. 88 (1987), 69–81.

[11] S. Bosch, W. Lütkebohmert and M. Raynaud, *Néron models*, Ergebnisse der Mathematik und Ihre Grenzgebiete 3. Folge, Band 21, Springer (1990).

[12] S. Chowla, *Proof of a conjecture of Julia Robinson*, Norske Vid. Selsk. Forh. (Trondheim) 34 (1961), 100–101.

[13] R. Coleman, *Effective Chabauty*, Duke Math. J. 52 (1985), 765–770.

[14] G. Cornell, J.H. Silverman, *Arithmetic geometry*, Springer 1986.

[15] M. Coppens, *A remark on curves covered by coverings*, Preprint.

[16] O. Debarre and R. Fahlaoui, *Abelian varieties in $W_d^r(C)$ and rational points on algebraic curves*, To appear in Compos. Math. (1993).

[17] O. Debarre and M.J. Klassen, *Points of low degree on smooth plane curves*, Preprint 1993.

[18] J.H. Evertse, *On sums of S-units and linear recurrences*, Compos. Math. 53 (1984), 225–244.

[19] J.H. Evertse , *On equations in S-units and the Thue-Mahler equation*, Invent. Math. 75 (1984), 561–584.

[20] J.H. Evertse, K. Győry, C.L. Stewart and R. Tijdeman, *S-unit equations and their applications*, In: New Advances in Transcendence Theory, Cambridge Univ. Press, 1988, pp. 110–174.

[21] G. Faltings, *Endlichkeitssätze für abelsche Varietäten über Zahlkörpern*, Inv. Math. 73 (1983), 349–366.

[22] G. Faltings, *Diophantine approximation on abelian varieties*, Annals of Mathematics 133 (1991), 549–576.

[23] G. Faltings, *The general case of S. Lang's conjecture*, Manuscript, 1991, 14 pages.

[24] G. Faltings and G. Wüstholz, *Rational points*, Seminar Bonn/Wuppertal 1983/84, Aspetcs Math. E6, Vieweg 1984.

[25] A. Grothendieck and J. Dieudonné, *Eléments de Géométrie Algébrique, Ch. I, II, III, IV*, Publications Mathématiques de l'I.H.E.S. 4, 8, 11, 17, 20, 24, 28, 32.

[26] G.H. Hardy & E.M. Wright, *An Introduction to the Theory of Numbers*, Oxford Univ. Press, 4th ed., 1960.

[27] R. Hartshorne, *Algebraic geometry*, Graduate Texts in Mathematics 52, Springer 1977.

[28] R. Hartshorne, *Ample subvarieties of algebraic varieties*, Lecture Notes in Mathematics 156, Springer 1970.

[29] M. Hindry, *Autour d'une conjecture de Serge Lang*, Inv. Math. 94 (1988), 575–603.

[30] M. Hindry, *Sur les conjectures de Mordell et Lang (d'après Vojta, Faltings et Bombieri)*, Astérisque 209, Société Mathématique de France (1992).

[31] R. Kannan, *Algorithmic geometry of numbers*, Ann. Rev. Comput. Sci. 1987, 2: 231-267.

[32] S. Kawamata, *On Bloch's conjecture*, Inv. Math. 57 (1980), 97–100.

[33] L. Lafforgue, *Faltings's "Diophantine approximation on abelian varieties"*, manuscript, not published.

[34] L. Lafforgue, *Une version en Géométrie Diophantienne du "Lemme de l'Indice"*, preprint LMENS–90–6 (1990), Ecole Normale Supérieure, 45 Rue d'Ulm, Paris.

[35] L. Lafforgue, *Le "Lemme de l'Indice" dans le Cas des Polynômes*, preprint LMENS–90–7 (1990), Ecole Normale Supérieure, 45 Rue d'Ulm, Paris.

[36] S. Lang, *Fundamentals of diophantine geometry*, Springer 1983.

[37] S. Lang, *Integral points on curves*, Publ. Math. I.H.E.S. 6 (1960), 27–43.

[38] S. Lang, *Number theory III*, Encyclop. Math. Sc., Vol. 60 (1991), Springer-Verlag.

[39] S. Lang, *Hyperbolic and diophantine analysis*, Bull. A.M.S. 14 (1986), 159–205.

[40] S. Lang, *Abelian varieties*, Intersc. Tracts Math. 7, Intersc. Publ. (1959).

[41] M. Laurent, *Equations diophantiennes exponentielles*, Inv. Math. 78 (1984), 299–327

[42] D.H. Lehmer, *On the diophantine equation* $x^3 + y^3 + z^3 = 1$, J. London Math.Soc. 31 (1956), 275–280.

[43] A.K. Lenstra, H.W. Lenstra Jr. and L. Lovász, *Factoring polynomials with rational coefficients*, Math. Ann. 261 (1982), 515–534.

[44] K. Mahler, *Über die rational Punkte auf Kurven vom Geslecht Eins*, Journ. reine angew. Math. (Crelle) 170 (1934), 168–178.

[45] Yu. I. Manin, *Rational points of algebraic curves over number fields*, Izv. Akad. Nauk 27 (1963), 1395–1440 [AMS Translat. 50 (1966), 189–234].

[46] D. Masser, *Elliptic Functions and Transcendence*, Lecture Notes in Math. 437, Springer.

[47] H. Matsumura, *Commutative algebra*, W.A. Benjamin Co., New York (1970).

[48] H. Minkowski, *Geometrie der Zahlen*, New York, Chelsea (1953).

[49] L.J. Mordell, *Diophantine Equations*, Academic Press, 1969.

[50] L. Moret-Bailly, *Pinceaux de variétés abéliennes*, Astérisque 129, Société Mathématique de France (1985).

[51] S. Mori and S. Mukai, *The Uniruledness of the moduli space of curves of genus 11*, Appendix: Mumford's theorem on curves on K3 surfaces. Algebraic Geometry, Tokyo/Kyoto 1982 (Ed. M. Raynaud and T. Shioda). Lecture Notes in Math. 1016, pp. 334–353, Appendix pp.351–353.

[52] D. Mumford, *Abelian varieties*, Tata Inst. of Fund. Res. and Oxford Univ. Press 1974.

[53] D. Mumford, *A remark on Mordell's conjecture*, Amer. J. Math., **87** (1965), 1007–1016.

[54] A. Neeman, *Weierstrass points in characteristic p*, Inv. Math. 75 (1984), 359–376.

[55] J. Noguchi, *A higher dimensional analogue of Mordell's conjecture over function fields*, Math. Ann. 258 (1981), 207–212.

[56] J. Noguchi, *Lemma on logarithmic derivatives and holomorphic curves in algebraic varieties*, Nagoya Math. J. 83 (1981), 213–233.

[57] T. Ochiai, *On holomorphic curves in algebraic varieties with ample irregularity*, Invent. Math. 43 (1977), 83–96.

[58] O. Perron, *Die Lehre von den Kettenbrüchen*, Teubner, 3. Auflage, 1954.

[59] A.J. v.d. Poorten and H.P. Schlickewei, *Additive relations in fields*, J. Austral. Math. Soc., Series A, 51 (1991), 154–170.

[60] M. Raynaud, *Courbes sur une variété abélienne et points de torsion*, Invent. Math. 71 (1983), 207-233.

[61] K.F. Roth, *Rational approximations to algebraic numbers*, Mathematika **2**, 1–20, Corr. 168 (1955).

[62] P. Samuel, *Théorie algébrique des nombres*, Collection Méthodes, Hermann (1967).

[63] H.P. Schlickewei, *An explicit upper bound for the number of solutions of the S-unit equations*, J. reine angew. Math. 406 (1990), 109–120.

[64] H.P. Schlickewei, *The p-adic Thue-Siegel-Roth-Schmidt theorem*, Arch. Math. 29 (1977), 267-270.

[65] W.M. Schmidt, *Diophantine Approximation*, Lecture Notes in Math. 785.

[66] W.M. Schmidt, *The subspace theorem in diophantine approximations*, Compos. Math. 69 (1989), 121–173.

[67] W.M. Schmidt, *Diophantine Approximations and Diophantine Equations*, Lecture Notes in Math. 1467.

[68] W.M. Schmidt, *Simultaneous approximation to algebraic numbers by rationals*, Acta Math. 125 (1970), 189-201.

[69] W.M. Schmidt, *Norm form equations*, Annals of Math. 96 (1972), 526–551.

[70] J-P. Serre, *Lectures on the Mordell-Weil theorem*, Asp. Math. E15, Vieweg 1989.

[71] J-P. Serre, *Groupes Algébriques et Corps de Classes*, 2ème édition, Hermann, Paris.

[72] T.N. Shorey and R. Tijdeman, *Exponential Diophantine Equations*, Cambridge Univ. Press, 1986.

[73] C.L. Siegel, *Lectures on the Geometry of Numbers*, Springer Verlag, 1989.

[74] C.L. Siegel, *Über einige Anwendungen Diophantischer Approximationen*, Abh. Preuss. Akad. Wissensch. Phys. Math. Kl. (1929), 41–69.

[75] J. H. Silverman, *Integral points on abelian surfaces are widely spaced*, Compos. Math. 61 (1987), 253–266.

[76] C. Soulé et al., *Lectures on Arakelov geometry*, Cambridge studies in advanced mathematics 33, Cambridge University Press 1992.

[77] N. Tzanakis and B.M.M. de Weger, *On the practical solution of the Thue equation*, J. Number Th. 31 (1989), 99–132.

[78] N. Tzanakis and B.M.M. de Weger, *How to solve explicitly a Thue-Mahler equation*, Memorandum No. 905, Univ. Twente, Nov. 1990.

[79] K. Ueno, *Classification theory of algebraic varieties and compact complex spaces*, Lecture Notes in Math. 439.

[80] P. Vojta, *Diophantine approximations and value distribution theory*, Lecture Notes in Math. 1239.

[81] P. Vojta, *Applications of arithmetic algebraic geometry to Diophantine approximations*, Lectures given at a CIME summer school in Trento, Italy (1991), Preprint, to appear in a Lecture Notes in Mathematics.

[82] P. Vojta, *A refinement of Schmidt's Subspace Theorem*, Am. J. Math. 111 (1989), 489–518.

[83] P. Vojta, *Integral points on subvarieties of semi-abelian varieties*, Manuscript December 1991, 26 pp.

[84] P. Vojta, *Arithmetic discriminants and quadratic points on curves*, In: Arithmetic Algebraic Geometry. Eds. G. van der Geer, F. Oort, J. Steenbrink. Progress in Math. 89 (1991).

[85] B.L. van der Waerden, *Geometry and Algebra in Ancient Civilizations*, Springer, 1983.

[86] B.M.M. de Weger, *Algorithms for Diophantine Equations*, C.W.I. Tract No. 65, Centr. Math. Comp. Sci., Amsterdam, 1989.

[87] A. Weil, *Basic number theory*, Springer 1967.

[88] A. Weil, *Variétés Abéliennes et Courbes Algébriques*, Hermann, Paris (1948).

[89] E. Wirsing, *On approximations of algebraic numbers by algebraic numbers of bounded degree*, Proc. Symp. Pure Math. **20**, Number Theory Institute, D.J. Lewis ed., AMS, Providence R.I., 1971, pp. 213–247.

Printing and Binding: WB-Druck, Rieden/Allgäu